EARTHQUAKES AND ANIMALS

From Folk Legends to Science

EARTHQUAKES AND ANIMALS

From Folk Legends to Science

Motoji Ikeya
Osaka University, Japan

 **World Scientific**

NEW JERSEY · LONDON · SINGAPORE · BEIJING · SHANGHAI · HONG KONG · TAIPEI · CHENNAI

Published by

World Scientific Publishing Co. Pte. Ltd.

5 Toh Tuck Link, Singapore 596224

USA office: Suite 202, 1060 Main Street, River Edge, NJ 07661

UK office: 57 Shelton Street, Covent Garden, London WC2H 9HE

British Library Cataloguing-in-Publication Data
A catalogue record for this book is available from the British Library.

Front cover:

A woodblock print of an earthquake catfish, Namazu-e, depicting a well-known Japanese legend about the revenge of local people on the catfish for causing an earthquake that took the lives of a number of local beauties. The catfish happily remarks that it will tremble again if more beauties appear. The same print appears on the frontispiece. (Courtesy of the Library of the Earthquake Research Institute, University of Tokyo.)

Cover design: Briar Whitehead

Kobe after the 1995 Earthquake. Photo credit (cover and title page): Dr Roger Hutchinson.

ISBN 981-238-591-6

Printed in Singapore by World Scientific Printers (S) Pte Ltd

A tornado-type cloud that appeared the evening before the Kobe Earthquake.

A woodblock print, Namazu-e (Earthquake catfish). An earthquake catfish shot with an arrow by the god Kashima commits *hara-kiri* in apology while gold coin boxes spill out of its belly in compensation. The vague shadows at the upper left represent people killed by the earthquake (ERI Library, Univ. Tokyo).

A photograph of *Earthquake light* (EQL) taken at the time of the Kobe Earthquake (Chapters 2 and 7; Wadatsumi, 1995).

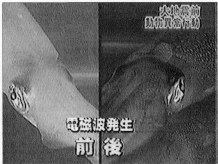

Left: *Earthworms* surfacing in large numbers at Kaoshung Elementary School, Taiwan before a large aftershock (Photo: Mr Chen). **Right: *Color change in a squid*** before (left) and after (right) stimulation by electric pulses (Chapter 4; TBS TV program, *Fantastic Animals*).

Experiments reproducing "earthquake clouds" (**EQCs**). (Upper left): Intense electric fields in a supercooled atmosphere at the electrodes generate clouds. (Upper right): An upward plume like a tornado or pillar cloud forms between the tip of a needle electrode (at the base of the plume) and a grounded metal sphere. (Chapter 7).

Bars on a rice leaf. These bars of discoloration, legendary precursors of earthquakes, were induced by electric pulses (Chapter 6).

(Above): Nails about to drop from a magnet on introduction of an electric field, reproducing an event known to have occurred two hours before the Ansei-Edo Earthquake in 1855 (Chapter 9).

(Right): Disturbance on a TV screen before the Kobe Earthquake. The first frame is normal but a rainbow-colored band and speckling appear in the second (Chapter 9). Professor S. Katoh appearing in the NHK TV program, *Humanity Lectures.* (From the book, *Why Do Animals Behave Unusually Before Earthquakes,* M. Ikeya; NHK, Tokyo.)

(Left): **Simulation of "The Moses phenomenon".** An unusual event, in which a Turkish fishing boat touched the sea floor between walls of water during an earthquake, was simulated in a narrow channel waterflow experiment using a red paper boat (Chapter 8).

Preface

More than 6400 people died in the Kobe Earthquake on January 17, 1995. At the time I was living 40 km away from the epicenter and I felt helpless thinking of the dead and injured. Like everyone else I heard the stories of trapped people urging family members to abandon their rescue attempts before they lost their own lives in the spreading fires. Some students left university to do voluntary work, but I was unable to relinquish my duties.

Two to three million people died in earthquakes during the 20th century. More than 240,000 perished in the Tangshan Earthquake in China and 20,000 in the Izmit and Indian earthquakes. I longed for some way to forewarn of these tragedies and so reduce the toll of injury and death but there were high levels of skepticism in the scientific community about short-term earthquake prediction and most had left the field to study earthquake mechanisms.

I have been involved for 30 years in an interdisciplinary field between solid state physics, geology and anthropology. My specialty has been dating, based on detection of accumulated unpaired electron spins produced by natural radiation in various materials, including gouge in geological faults (summarized in my book *New Applications of Electron Spin Resonance*). Obviously, knowledge of the Earth's past isn't of much use in reducing earthquake casualties, but I wondered if I might have a contribution to make based on my years of working between several disciplines, that is, as an amateur having a scientific background making no claims to any professional expertise in earthquake prediction.

I began to apply my experience to what I was hearing from witnesses of the Kobe Earthquake. Some professors and students saw earthquake light (EQL) just before the Kobe Earthquake. I started analysis of the light based on my *electromagnetic (EM) model of a geological fault*. Electric discharges ahead of the main shock seemed to explain EQL as an atmospheric dark glow.

Earthquake precursor reports, which were retrospectively collected upon request through mass media and published as *1519 Statements on Precursors before the Kobe Earthquake* by Wadatsumi (1995), and also collected and analyzed by a subcommittee of the Kansai Science Forum*, echoed the stories common in our folklore and in the book *Earthquake Catfish* by K. Musha. We also collected stories of pre-earthquake phenomena before the earthquakes in

Taiwan and Izmit (Ulusoy and Ikeya, 2001) in 1999, from refugees from the epicenter area. I found the stories were also similar to the legends and to a database classified by Rikitake (1976, 2001) and discussed by Tributsch (1982).

I don't deny that there are difficulties with the lay reports but there were trends that seemed to me could be tested scientifically. I wondered if the unusual animal behavior commonly mentioned in reports might be produced by the same electric pulses that created earthquake light and decided to try the principle on animals and plants but at much lower intensities, again using the EM model. I found that fish aligned themselves in formation (as the reports said), mimosa (the sensitive plant) closed its leaves and bowed and animals behaved much as the legends described. Mysterious earthquake precursor phenomena in legends and puzzling modern reports on malfunctioning home electric appliances, which science considers anecdotal, were simulated in EM experiments conducted in the laboratory.

The EM model of a fault was found to be an appropriate explanation—not just at the point of fault movement but also at the microfracture level—of the means by which EM waves could travel to the surface and have the effects that were being produced in the laboratory and reported in precursor statements. There are a number of mechanisms—relating to increasing pressure and diffusion of water into microfractures—by which an increasing amount of separated charge can be produced in underground rock. Any change in position of charge or in charge density can generate EM waves in a wide range of frequencies. Low frequency (ULF) waves (unlike higher frequency waves) can travel to the surface without a conductive path, and, on the surface, are capable of creating intense electric fields and generating EM waves up to very high frequencies (VHF). In our laboratory experiments animals, plants and electronic objects were exposed to pulsed and static electric fields and their behavior compared with the electric field expected from the EM model.

We settled on the piezoelectricity (electric polarization) of quartz as the principal mechanism for the production of underground charge. Many earthquake-prone areas are granite or volcanic and, so, high in quartz, and the charge is produced as a result of stress on the quartz crystals in such rock. Piezocompensating charge (produced to cancel out piezoelectric charge) also adds to the increasing amount of charge, all of which, on movement, generate EM waves. In non-volcanic or non-granite regions other mechanisms of charge generation may be at work underground, though quartz grains are present in most rock, including sedimentary rock, and only a tiny amount of orientation of the crystals is necessary for a considerable amount of charge to be produced. There will be regions of the earth where tectonic action is so deep that no ULF waves can reach the surface, in which case there will be no EM precursors.

However, when I started to write up these findings for publication in scientific journals, I met outright rejection by referees on grounds that anecdotal statements by lay citizens were not a legitimate subject of scientific research and that, in any case, prediction was not possible in principle. Some students and friends regarded the work as "high risk and no return" and advised me to quit. I alternated between excitement and despair: excitement at the outcome of the experiments and the development of a new theory, and despair at the resistance I met in the scientific community. I felt like *Alice in Wonderland* caught between fantasy and reality. No wonder most scientists avoid investigations on the boundary of science and legend. I could not get any research grant and struggled in a scientific wilderness. However, 40 related papers were still published in scientific journals over eight years and two books for the Japanese market, *Why Do Animals Behave Unusually?* and *Precursors of Large Earthquakes.*

The general public and the mass media were sympathetic. They expected me—a scientifically trained person—to look for some scientific basis to the stories of pre-earthquake phenomena and to develop a method of prediction if possible. A few friendly seismologists constructively suggested that short-term forewarning might be possible if limited to unusual animal behaviour produced by electromagnetic fields, and invited us to join discussions.

I am not arguing for earthquake prediction using the animal precursors described in *Earthquakes and Animals,* if, by prediction, we are meaning an exact forecast of time, epicenter and magnitude of an earthquake. In that case no prediction is possible for any fracture phenomena. Besides, the research is too young and animal behaviour relative and sometimes idiosyncratic. But I *am* arguing that there is a scientific basis to many of the legendary and reported precursors. I am also arguing for the use of some animals e.g. the catfish, *along with electronic observation of other SEMS anomalies,* in general forecasting of large earthquakes. That is, I believe that an informed interpretation of such amalgamated data should permit a rough estimate to be made of the likely region, time and magnitude of a large quake. The region and magnitude can be derived from the extent and intensity of known animal responses and other SEMS activity, though time might remain rather vaguely defined e.g. sometime within a few weeks.

A population generally educated about earthquake precursors is a prepared population. People usually need to be prodded into earthquake preparedness and observation of a range of precursor anomalies (from the odd behaviour of pet animals to malfunctioning home appliances and cell phones) over a relatively short period, helps to do that.

Of course, nobody can stop an impending large earthquake, but people can at least take simple precautions like fastening furniture to the wall and checking emergency items. I do hope that precursor phenomena caused by SEMS will be taken seriously by people living in earthquake-prone areas. If there is an earthquake, lives and property will be saved; if there is no earthquake we can simply be glad.

I believe that science is now mature enough that scientists from different specialties working cooperatively, and also working with laymen, might be able to reduce earthquake casualties. In this research I have moved too far from my field of specialization and I appreciate this will invite criticism by specialists. The research may also look elementary to specialists, but, again, this is inevitable in such interdisciplinary work. I welcome constructive criticism, comments and more elegant experiments and theories. If I have made mistakes, I still consider some loss of reputation a small price to pay for work that may help save lives.

I wrote this book to show that many ancient legends warning of impending quakes have a scientific basis; they are not merely superstitions but can be explained as rare natural electromagnetic phenomena that may forewarn of earthquakes and therefore help minimize casualties. The experiments described in this book are simple enough: educational for the general public, and may be demonstrated in high school classes, general lectures and science shows, as I am doing regularly.

Chapters 1 and 2, respectively, of *Earthquakes and Animals* describe legends and recent retrospective reports of earthquake precursors. Chapter 3 gives elementary descriptions of earthquakes and electromagnetism and may be skipped over by scientists; Chapter 4 describes experiments on animals showing behavioral responses to electromagnetic pulses and Chapter 5 describes a rock compression experiment and animal responses to EM waves produced by fracture. Chapter 6 discusses plant reactions to EM exposure; Chapter 7, unusual atmospheric phenomena such as earthquake light, clouds, fogs, rainbows and unusual formations in the sky. Chapters 8 and 9 offer EM explanations for other unusual earthquake-associated phenomena that have been left unexplained or for which alternative explanations have been offered, and also describe laboratory simulation of reports of malfunctioning domestic electric appliances. In the absence of any sure-fire methods of earthquake prediction, Chapter 10 presents automated monitoring of unusual animal behavior, in particular the catfish, as an experimental forecasting model. Chapter 11 generally surveys the seismo-electromagnetic signals (SEMS) research field, which has made significant advances since the Kobe Earthquake. Chapter 12 summarizes the book, and two appendices present results of a questionnaire on precursor phenomena and a short section on disaster prevention.

Throughout the research, I was aware of those millions who have been killed by earthquakes, and their families, and carried the hope that the results would help reduce the death toll. I also enjoyed the science. Again I thank the editors of *Earthquakes and Animals* and all those listed in the acknowledgements—not forgetting those members of the general population who encouraged me to complete the work on their behalf.

January, 2004, Motoji Ikeya

* The Third Subcommittee on the Utilization of the Information on Earthquake Precursors, chaired by N. Kumagai, former President and Professor Emeritus of Osaka University.

Acknowledgements

My thanks to Dr Neil Whitehead of DSIR, New Zealand, and Ms Briar Whitehead; Dr Himansu Kr Kundu of the Geological Survey of India, Calcutta, and Dr Yoshiko (Okada) Ikeya, for their editing and proof-reading.

Figures and illustrations were drawn by Ms Nami Matsumura, Mr A. Hoch, Dr T. Takaki, Mr H. Asahara and Y. Emoto. Ms Briar Whitehead prepared photos, figures and manuscript for press.

My thanks to my coworkers: Associate Professors C. Yamanaka and C. Huang, and Professors M. Ohta, N. Ohtani, N. Kajiwara, Y. Gondou, K. Wadatsumi, K. Nagai and K Yagi; also to members of my laboratory for studying electromagnetic earthquake precursor phenomena after the Kobe Earthquake and to members of the Kansai Science Forum, especially Professor T. Kinoshita, for their discussions. Also to Associate Professor Ulku Ulusoy at Hacettepe University, Dr H. K. Kundu (above) and Dr T. Shih of the Geological Survey of Taiwan, who surveyed precursor phenomena associated with the Izmit/Turkey, Gujarat/India and Taiwan-921 earthquakes, respectively.

I warmly acknowledge the comments, help and encouragement I received from Professor S. Uyeda, N. Nagao, M. Kumazawa, H. Mitsudo and the late Professor R. Uyeda, throughout the research. My thanks too to the Kansai Energy Recycling Foundation (KERF) and Frontier Research Center (FRC) of Osaka University for their support in the interests of disaster prevention.

Thanks also to publishers and copyright holders for kind permissions to use the following figures:

Am. Association for the Advancement of Science (Science) Fig. 3.5; Am. Geophys. Union Fig.11.4, Fig.11.11; Asahi Newspaper Co. Fig. 4.1; Elsevier Fig.11.8 (J. Geodynamics); Goma Press Fig. 6.4; Jpn Academy of Science Figs 6.2 11.6; Kansai Science Forum Fig. 2.15; Earthquake Research Institute Library, University of Tokyo, for all namazu-e: front cover, front color plates (fcp), title page of Ch 1, Fig.1.5; McMillan Publisher Ltd (Nature) Fig. 7.8; NGS Publisher Figs 1.2; NHK Publisher and NHK Broadcasting Co. fcp, Fig.7.3 and figures from the book, *Why do Animals Behave Unusually Before Earthquakes,* M. Ikeya, NHK, Tokyo; Sankou Publisher Fig. 7.9; Kluwer Academic Publishers Fig. 11.1 (Surveys in Geophysics 18 (1997), p441-476, f.1a, p443, f.1b, p444); Seisyn Press Fig. 1.3; Society of Atmospheric Electricity Fig.10.5; TBS (*Amazing Animals*) fcp (squids) Figs 4.11, 4.17, 4.18; Prof. K. Wadatsumi, photographs in *Precursor Statements 1519* (Tokyo Publisher) fcp (EQC and EQL), and Figs 2.1, 2.2, 2.3.

Motoji Ikeya:

Professor of Graduate School of Science, Osaka University (Department of Physics since 1987 and of Earth and Space Science since its foundation in 1991), chair of the Quantum Geophysics Laboratory. Author of more than 300 published papers. Majored in Electronics and then Nuclear Engineering at Osaka University, worked at Nagoya and Yamaguchi Universities, as a research associate at UNC-CH, USA and as a research fellow of the Alexander von Humboldt Foundation at the University of Stuttgart, Germany. Recipient of the Asahi Newspaper Grant for Encouragement of Science (1981) and the 4th Osaka Science Prize in 1986.

Major fields of specialization: Quantum geophysics—interdisciplinary research into Electron Spin Resonance (ESR) for dating of geological and archaeological materials and, in the future, icy planetary bodies. Also radiation dosimetry and assessment of the paleo-environment. Earthquake precursor studies after the Kobe Earthquake in 1995.

Main publications: *New Applications of Electron Spin Resonance—Dating, Dosimetry and Microscopy,* World Scientific (1993; 2002). Books in Japanese are *ESR Dating,* Ionics (1987); *ESR Microscopes,* Springer-Tokyo (1992); *Why Do Animals Behave Unusually?—Birth of Electromagnetic Seismology,* NHK Publisher (1998) also translated by C. Huang into Chinese (Suichen Press, 2000); *Precursors of Large Earthquakes,* Seisyun Press (2000).

Contents

Legends of Unusual Phenomena Before Earthquakes
— Wisdom or Superstition?

Three catfish that caused three major earthquakes in the Edo era were captured and taken to an eel broiler by the god, Kashima. Carpenters and fire fighters who profited by the earthquakes ask the god for mercy (Earthquake Research Institute (ERI) Library, University of Tokyo).

1.1 Introduction

In Asian countries where earthquakes occur frequently, often with many casualties, an abundant folklore of legends and proverbs about earthquake precursor phenomena has grown up. The old Japanese legend of the earthquake catfish—known all over the world—is one of them. It tells of a large catfish living underground that causes earthquakes whenever it moves. The legends probably derive from the unusual behavior—violent jumping and twisting movements—that has been observed in catfish before earthquakes.

There are numerous legends about fogs, unusual cloud formations, lights, and unusual animal behavior before earthquakes. But the stories are not only legendary. Fogs, unusual cloud formations, lights, and unusual animal behavior, from a week to a few minutes before major earthquakes, have been widely reported in the modern world.

It is possible that these legendary and reported unusual events may be found to have a scientific basis, just as anthropology has often found grains of truth in folk tales and taboos. However, most retrospectively reported pre-earthquake phenomena fall between scientific disciplines and scientists are understandably hesitant to undertake research outside or on the boundaries of their specialties. They are also reluctant to leave the hard sciences to investigate lay observations. But unless they do so the legends and reports will remain only legends and reports, even though there is strong public interest in what science has to say.

There were some early attempts at scientific analysis in the 1930s. Musha (1931) studied historical materials and collected statements from witnesses of the earthquake at Izu, near Tokyo. A Japanese X-ray physicist and essayist, Terada (1931), one of the founders of the Earthquake Research Institute (ERI), University of Tokyo, wrote about earthquake light (EQL):

The present writer was of the opinion that the phenomena may well be worth a serious study and at least cannot be discarded as trivial without scrutiny.

Fireballs were once said to have accompanied ghosts, devils, thunder and earthquakes. Some people tried to explain them anecdotally as unidentified flying objects (UFO) carrying extraterrestrials (ET). But science rehabilitated fireballs when Nobel prize-winning Russian scientist Kapitza suggested that fireballs in nature were produced by natural radiowaves (Kapitza, 1955), and Ohtsuki and Ofuruton (1991) successfully reproduced a fireball in a laboratory. It is not unreasonable to believe that legendary earthquake precursor phenomena might be similarly explained scientifically—but only after serious and rigorous scientific evaluation.

Stories of mysterious earthquake precursors have been described in detail in many books, the most well-known of them by Musha (1957,1995) in Japanese and by Rikitake (1976, 2001) and Tributsch (1982) in English. Several books have been published in Chinese, but appear not to have been translated into English. Unfortunately, publications on pre-earthquake phenomena have largely been motivated by people's eagerness to find predictors, and only Terada and Rikitake have undertaken serious scientific studies. Since other books comprehensively discuss legends associated with major earthquakes outside Japan, this chapter will restrict itself mainly to Japanese legends and proverbs.

1.2 Precursor phenomena in the sky, the atmosphere and on land

1.2.1 Earthquake light (EQL)

Well before the advent of electricity mysterious lights were often reported to have accompanied earthquakes. Daigo (1985) quotes some old Japanese sayings,

A red light that spreads like a big fire over an usually dark area at night is a sign of a big earthquake or tsunami close to that area.
Red clouds at nine o'clock on a dark night warn of an earthquake.

Musha (1957) refers to the following stories from old books on earthquake light in his book *Earthquake Catfish*.

Sparks like flashlights: An old record on EQL mentions several intensely brilliant flashes of light at the time of the Mutsu Earthquake in 869 AD. People screamed and were thrown to the ground or crushed under collapsing houses. They reported hearing sounds like thunder just before the quake and big waves (tsunami) destroyed the town.

A red light like fire in the sky: Before a large earthquake at a town on the coast of the Japan Sea in 1751,

People went out to sea in the evening to fish. Thirty to 40 km off the coast they noticed a red light in the sky in the direction of their town. Fearing it was on fire they returned to land as fast as possible but saw no evidence of fire and found the town was safe. Puzzled, but pleased, they drank a cup of tea. At midnight, there was the sound of a loud bang like a cannon and they knew no more. A mountain close to the coast split into two and sank into the sea. There was only one survivor, a housewife who was rescued from the sea, and who told the story.

Fireball, flames and other radiation: Many people observed strange lightning at the time of the Zenkoji Earthquake in 1847. In addition to a pillar of fire (or fire rod) and a trumpet of fire extending upward from the source of light, luminous variously shaped, moving objects like fireballs (ball lightning) were reported. On the ground dried grasses were burned in the flames.

Skeptical scientists attribute EQL to meteorite showers or discharges from electric power lines. However, photographs were taken of luminous phenomena in restricted areas near the epicenter of the Matsushiro Earthquake swarm in 1965 - 1967 (See Figure 1:1, Yasui, 1968). After that EQL became a subject of scientific study (Derr, 1973). Thunder-type EQL was also reported to have occurred just before the earthquakes, and luminescence has been observed at the time of laboratory rock rupture. Hence, EQL may be considered a coseismic as well as a preseismic phenomenon. Possible mechanisms of EQL based on atmospheric discharges are described in Chapter 7.

1.2.2 Earthquake clouds (EQCs) and fogs (EQFs)

Peculiar or special types of fogs are said to appear before earthquakes and there are a number of Japanese proverbs on earthquake clouds.

When clouds are close to the ground there will be rain or an earthquake.

An earthquake will occur when clouds move eastward or when a long black striped dragon-like cloud floats horizontally.

A mackerel (or cirrus) sky precedes an earthquake.

Fine weather fogs shrouding the mountains portend an earthquake.

The clouds stop moving, the wind stops, the scenery blurs and the earth rumbles before an earthquake.

Figure 1.1 Night photographs of Earthquake light (EQL) during the Matsushiro earthquake swarm in 1967 (Yasui, 1968).

Clouds like snakes or dragons surround the morning sun before earthquakes.

An old manuscript published in 1759 describes 20 types of clouds and accompanying weather changes said to precede earthquakes; some of them are shown in Figure 1.2.

Another old book published in 1830 tells the story of the Sado Earthquake (M6.6) in 1802 (Musha, 1957). Sado Island in the Japan Sea was once famous for its gold mines, and the miners used to leave the mine whenever fogs appeared, interpreting them as a pre-earthquake phenomenon. The fogs were thought to have been formed by the emanation of "earth air" (gas).

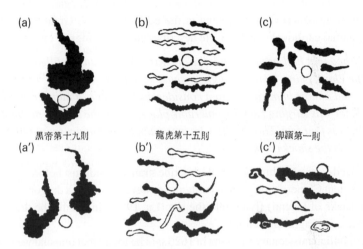

Figure 1.2 A drawing of clouds in the early morning and evening in a book published in 1759, indicating (a) heavy rains or storms that day or the following day, or if not, a big earthquake on the third day; (b) thunder within four hours, or, if the clouds appear between September and April, a big earthquake and (c) a terrible thunderstorm or a big earthquake. (a'), (b') and (c') show changes in shape. (Kamei, 2000).

A boatman noticed strange weather and clouds he had never seen before, close to the ground. There was no wind, and it did not rain. Soon after there was an earthquake.

Rikitake (1976), a geophysicist famous for his dynamo model of the Earth's magnetic field, referred to the above and other stories in which an old gate-keeper predicted a big earthquake because of fogs and unusual cloud formations over half the sky. Rikitake was cynical. "Ridiculously enough," he wrote, "people said the stars seemed to be very close." His conclusion was that the legends were probably fiction and the unusual weather mere coincidence. If the prestigious professor had not taken such a view he would undoubtedly have been ostracized by his scientific community, but our experiments give scientific support to these anecdotes.

The oldest record of earthquake clouds is found in Egyptian mythology, but Aristotle also discussed earthquake gas. In these accounts local dense fogs form at the epicenter area, but the sky is clear just a few meters from the fog. The accounts are similar in both Japan and Egypt. Tributsch (1982) referred to a similar story by Alexander von Humboldt in South America: fogs covered the epicenter area before a large earthquake and miners working there—like the miners on Sado Island—left quickly warning of a coming quake because of the sudden fog.

The Bulletin of the Seismological Society of America (Jaggar, 1923) carries the report of an American vacationing at the epicenter of the Great Kanto (Tokyo-Yokohama) earthquake in 1923. He describes coseismic fogs or clouds.

> *Although it had been raining, the air was almost instantly filled with dust, the cloud being heavy enough to cause temporary darkness. The sea re-ceded almost immediately to a considerable distance, a quarter of a mile. Water was surging from innumerable places along the seabed probably as high as four feet. The shocks were incessant, the miniature fountains spurt-ing as the shocks came.*

Thousand-year-old Chinese records describe strange clouds just before earthquakes. The ancient Chinese thought that earthquakes were caused by a dragon (the winged giant snake of legend) flying into the sky (Lu, 1981). In poems the dragon is de-scribed as a dragon-like cloud.

An Italian missionary in China in 1626 said he found that unusual weather: fogs and clouds, storms and sudden changes of atmospheric temperature preceded earth-quakes (*Interpretation of Earthquakes*, Yue and Lu, 1988).

A mayor of Nara City who gave credence to the precursor legends successfully predicted the Fukui (M7.1) and Izu-Ohshima (M7.0) earthquakes. He named the clouds "earthquake clouds ", published a book in Japanese (Kagita, 1983) and orga-nized a group of cloud watchers to help predict earthquakes. However, similar clouds also form in areas where there are no major earthquakes.

1.2.3 Sky, sun, moon, stars and rainbows

Sky: As in the old Mother Goose rhyme

> *Red sky at night, shepherd's delight.*
> *Red sky in the morning, shepherd's warning.*

A red sky in the evening indicates fine weather, but a red sky in the morning suggests rain. There is a scientific explanation. In Japan, the weather changes from west to east because of the region's prevailing mid-altitude westerlies. High cloud is usually associated with rain fronts. A red sky in the morning means the light of the rising sun is reflecting off the frontal cloud approaching from the west and there will be rain. (The sky is red because of the way light bends through the atmosphere.) Red sky in the evening means that the setting sun is reflecting off the frontal cloud that has just passed and there will probably be a period of calm weather before the next front approaches.

Not only the clouds, but the sky itself is said to look different from normal before earthquakes, particularly the reds and yellows of morning and evening skies.

An unusually red sky in the evening is an indicator of earthquakes.
An earthquake occurs when humid weather lasts a long time, the sky looks
*   yellow, and a red light like a big fire spreads into the night.*

Sun: The sun with an 'umbrella' (a halo) around it (from small ice particles) is a sign of rain in Japan. Tributsch (1982) referred to a Chilean proverb which goes,

When the sun has a ring around it, there will be rain showers or an
*   earthquake.*

Moon: Haloes round the moon are said to be a sign of rain or an earthquake. An elongated moon, a yellow moon and a moon with haloes of various colors are often said to be signs of an earthquake. A Japanese proverb on earthquakes says,

A yellow morning sun, a red moon and twinkling stars are precursors.

Stars: When even small stars twinkle vividly and intensely and appear to be very close to the earth, a major earthquake is said to be on the way. Mysterious appearances of the sun, moon and stars have been thought to be earthquake precursors in many countries since the time of the Romans (Tributsch, 1982). A Japanese proverb says,

If the stars seem unusually close, there will be an earthquake.

Rainbows: Old Japanese sayings talk of "Hinoko" (Fire-powder), a characteristically short and vertical rainbow between mountains, which they say presages a big earthquake.

A blurred, vertical rainbow is a sign of an earthquake.
White rainbows, three-colored black, yellow and violet rainbows, and
*   rainbows near the ground are signs of a coming quake.*

Table 2.4 (Chapter 2) summarizes some of the precursor phenomena reported in the sky and atmosphere. Their mechanisms are still not clear and none have been clearly linked with impending earthquakes or coseismic fault movement. They will be discussed further in Chapter 7.

1.2.4 Weather, water and sounds

(a) Hot and humid weather:

Proverbs linking weather and earthquakes say,

> *An earthquake will occur when fine weather lasts a long time without clouds and wind.*
> *A lengthy period of extraordinarily hot and humid weather precedes an earthquake.*

The diffusion of heat from an underground earthquake site is probably too small to explain temperature gain before earthquakes. It could be accounted for merely by increased humidity above the epicenter area due to random local greenhouse effects which prevent the release of infrared rays.

> *When the wind blows up from the ground, there will soon be an earthquake.*
> *No wind for more than 10 days is a sign of an earthquake.*
> *Windless high humidity with a red sky in the evening is an earthquake warning.*
> *No wind or clouds after a big earthquake mean there will be aftershocks.*
> *If the scenery is blurred and mountains look distorted, there will be an earthquake.*
> *When the soil in pottery ware dries up unusually quickly, there will be an earthquake (See Figure 1.3).*

(b) Water: changes in temperature, turbidity and level

Legends on earthquake precursors related with ground water, hot spring and seawater are abundant in earthquake-prone countries. The Japanese proverbs are:

Well and hot spring water:

Figure 1.3 *When the soil in Bonsai pots dries up unusually fast, there will be an earthquake* says a proverb. There are many proverbs about weather and big earthquakes. (Ikeya, 2000; Seisyun Press)

If well water becomes turbid for more than a week, there will be an earthquake.

A sudden drop or increase in the water level in a well is a sign of an earthquake.
When hot spring water becomes white and turbid, there will be a major earthquake.

River and seawater:

The flow of the River Seta reversed before an earthquake.
Sudden tidal withdrawal is a sign of a coming big quake.
High temperatures of seawater and a red sea are earthquake indicators.

Tsunamis can cause havoc in coastal areas (See Chapter 3 and Living God, next page). Tidal withdrawal and parting of the sea ("The Moses Phenomenon") are demonstrated in a model experiment in Chapter 8.

(c) Sounds of explosions

There are legends about low frequency sounds.

Low frequency sounds from the earth precede an earth tremor.
Loud bangs like a firing cannon or firework explosions portend quakes.

Sound velocity is only 340 m/s. Seismic waves, about 3.5 km/s, are ten times faster. In this case the sounds would only be heard after quakes, but people report noises—sometimes like explosions—before earth tremors.

1.3 Unusual behavior of animals and plants

1.3.1 Unusual animal behavior: myths or reality?

For centuries, the world over, there have been stories of uncharacteristic or unusual or even bizarre animal behavior before earthquakes, described in terms of excitability and panic. For thousands of years Oriental myths and legends have told stories of animals predicting earthquakes. This has been of interest to popular science and to the media, but has rarely attracted serious scientific study.

In August 1971 the State Seismological Bureau of China started to collect reports of unusual animal behavior for earthquake prediction purposes. Four years later, based on observations of unusual animal behavior and geophysical measurements, they successfully evacuated Haicheng city several hours before an earthquake (M7.3) on February 4, 1975. This earthquake caused considerable damage to existing structures and cultivated lands, and the successful evacuation was thought to have saved 100,000 lives. There were also reports of unusual animal behavior before the Tangshan Earthquake (M8.2) in 1976, but no warning was issued. There were 240,000 casualties.

Living God: a story of a Tsunami

Lafcadio Hearn (Yagumo Koizumi), a novelist who introduced traditional Japan to the English-speaking world, published an article, *Living God*, about an earthquake tsunami in the book *Gleaning in Buddha-Fields* in 1897. The following paragraphs are taken from the article:

> *The day had been oppressive; and in spite of a rising breeze there was still in the air that sort of heavy heat which, according to the experience of the Japanese peasant, at certain seasons precedes an earthquake. And presently an earthquake came.... He rose to his feet, and looked at the sea. It had darkened quite suddenly, and it was acting strangely. It seemed to be moving against the wind. It was running away from the land...*

Village people were running to the beach to watch the enormous ebb. *Hamaguchi* hurried with a torch to the fields, where hundreds of rice-stacks, representing most of his invested capital, stood awaiting transport. He set them alight and the villagers, thinking his house was on fire, ran to the top of the hill to extinguish it. He stopped them there and waited until all the villagers gathered.

> *"Kita (It's coming)" shouted the old man at the top of his voice, pointing to the open sea. "Say now if I be mad!" Through the twilight eastward all looked, and saw on the edge of the dusky horizon a long, lean, dim line like the shadowing of a coast where no coast ever was, a line that thickened as they gazed ...*

As the story goes, the villagers whose lives he saved thought he must have been a god and wanted to worship him. But he was no god. He worked for the local government, built a stopbank to protect the village from the next tsunami, went to work as a surveyor in America after his retirement and died in New York. The story was written up in a national elementary school fifth grade textbook, but was removed after the War.

The notion that unusual animal behavior can help people predict earthquakes is dismissed by most Western scientists who tend to put it in the same category as sightings of UFOs and the Loch Ness Monster. But it is scientifically well established that many animals have a perceptual range well exceeding humans'. Possibly, also, they may have a sensitivity to earthquake precursor signals that has not yet been understood scientifically.

1.3.2 Legends and proverbs about animals before earthquakes

The following is a list of proverbs of unusual animal behaviors before quakes (Daigo, 1985).

Mammals: Cats *leave home and dogs howl mournfully, refusing to eat.*
There is an exodus of rats and cats from the area.
Animals come out of burrows.
Bats fly and forage in winter months when they would normally be hibernating.

Birds: *Pheasants and chickens cackle and cheep during the night.*
Chickens fly onto roofs. Cocks crow and hens cluck.
The pheasants' morning song means fine weather, but night cries mean an earthquake.

Reptiles: *Snakes and frogs come out in winter.*
Snakes gather in bamboo bushes.

Fish: *Big catfish stir deep underground.*
Fish rise to the surface of the water.
Fishing catches decrease near the coast.
Deep-sea fish float to the surface.
Goldfish or minnows burrow into the sand at the bottom of their aquaria.
Carp and goldfish jump out of the water.
Big shoals of sardines swim up-river.
Dizzy octopuses float up to the surface.
Colonies of sea crabs walk onshore.

Insects: *Red dragonflies swarm in the same direction.*
Honeybees buzz and swarm out of their hives.
Centipedes are found dead in random places.

Worms: *Earthworms come out in winter.*

There are also modern reports of pre-earthquake behavior. Among them:

Cows: became unusually restless, snapped their halters and escaped.
Pigs: rooted at fences and bit each other.
Rats: 150 km from the epicenter of the Haicheng Earthquake were agitated and appeared dazed and disoriented before the quake.
Rats and birds: left their normal habitats or were seen moving in a large numbers as if migrating.
Seagulls: flew inland in large numbers.
Parakeets (budgerigars): cried, fluttered and fell in distress or died.
Tortoises: emerged from their burrows in winter.

Snakes: came up to the surface (and froze in winter temperatures) or gathered in bamboo groves in summer.

Fish: in ponds and tanks oriented themselves in the same direction.

G*oldfish:* leapt out of their glass bowls and made unusual sounds.

Skepticism and laughter greets the story of the "catfish that cried" before the Tangshan Earthquake. However, according to zoologists, fish may "speak". In fact, if Chinese catfish are targeted with electric pulses, they make a distinctive sound after each pulse.

1.3.3 Unusually active catfish and the absence of eels

Of all Japanese animals the one most famously linked to earthquakes is the catfish. Celebrated catfish stories were written up in the *Ansei Chronicle* after the Edo (Tokyo) Earthquake of 1855. In one of them a man said that all the eels seemed to have disappeared from a local river and all he could see was a number of violently thrashing catfish. According to the account his wife laughed at him when he said an earthquake must be coming, but his life was saved because of it. (Musha, 1957; Rikitake, 1976).

The story may mean that eels, being more sensitive to earthquake signals than catfish, had already left the scene (Figure 1.4). Or it might mean that the eels had already burrowed into the sand but the catfish had no place to hide. (See Chapter 4)

Figure 1.4 Did sensitive eels escape or hide before the Ansei Earthquake, leaving the less sensitive catfish behind? (A. Hochi).

1.3.4 Color woodblock-prints of catfish: Namazu-e

According to the Japanese legend, a big catfish living underground shakes itself and causes earth tremors. Color woodblocks—innovative printing technology at that time—called Earthquake Catfish Pictures, "namazu-*e*", were circulated after the Ansei Earthquake in 1855. One comical catfish woodblock is on the cover of this book. Many girls were killed by the Ansei Earthquake at New Yosiwara (notorious for public prostitution at that time). The catfish, captured by the local people, says "I am delighted the beauties (high class educated prostitutes) clambered on my back. If any more clamber over me, I may tremble again". Carpenters and fire-fighters, who profited by the earthquake, run to the site and beg for the life of the catfish.

The catfish on the first color plate is taken from a woodblock-print describing an earthquake catfish apologizing for the earthquake he caused and committing suicide, *hara-kiri*, after being shot by the god at the Kashima shrine. The front page of Chapter 1 shows a *namazu-e* print of three cute catfish taken to a broiler store by the same god. Again, carpenters are asking the god to save the lives of the catfish.

Figure 1.5 is another woodblock print depicting the era's Four Fearful Ones: earthquake, thunder, fire and old fathers in that order, all comically personified.

Sarcastic remarks were often made about those who gained from earthquakes and against governments that were either incompetent or unable to deal with the disaster. It was cynically believed earthquakes occurred to punish politicians and in some prints revolutionary catfish are helping innocent citizens from collapsed buildings (Rikitake, 2001).

Were there any legends linking earthquakes and catfish before the *Ansei Chronicle* accounts? Musha (1957) referred to a poem written in 1680 which linked a dragon-like earthquake cloud to a catfish. Hideyoshi Toyotomi, a famous Japanese commander, wrote to his followers to observe the behavior of catfish when a new castle was being built. The Fushimi Earthquake in 1596 (M 7.5) destroyed his castle. So the catfish legend is at least 400 years old.

Figure 1.5 From a woodblock describing a game played by the Four Fearful Ones, earthquake (catfish), thunder (a devil with horns), fire (with fiery hair) and a father still having prestige (ERI Library, University of Tokyo).

Similar legends in Asian countries of the Altaic language group—to which Japanese, Manchu, Mongolian, Tungusic and Turkish belong—speak of a big subterranean fish, frog, duck, turtle, bull or mammoth supporting the land, which moves violently producing earthquakes (Tributsch, 1982). Ancient peoples sought to predict earthquakes from observation of animal behaviors like those of the archetypal underground animal.

1.3.5 Do human beings feel anything before earthquakes?

Is it only animals that behave strangely before earthquakes? Do humans feel anything unusual before large earthquakes? If so, can they predict earthquakes? Again, in the *Ansei Chronicle*, a story is told of a person loudly warning of an impending earthquake two hours before it happened. Some proverbs suggest that some people might be capable of sensing earthquake precursors (Daigo, 1985).

> *Some people feel peculiar fatigue or suffer from headaches or hysteria and have difficulty in breathing before earthquakes.*

Some report nervous irritability, restlessness, nausea or a sensation of dizziness and loss of balance. People in Japan and China who claim to be especially sensitive, talk about body symptoms similar to those described in legends and proverbs. Some of these "sensitives", some of the elderly and persons suffering from some illnesses, also do not feel well ahead of bad weather. Are they reacting to something bad weather and quakes have in common?

1.3.6 Early flowering, re-flowering and sensitive plants (Mimosa)

There are many literary references describing early sprouting, early blooming and re-blooming of plants before earthquakes.

> *There will be a exceptional crop of pears and persimmons before an earthquake.*
> *Before a big earthquake, plants flower out of season and re-flower, and vegetables almost come to maturity.*
> *The sensitive plant, mimosa, closes its leaves and bows its stems before earthquakes.*

Reportedly some nonseasonal flowers, shrubs and trees bloom in winter before earthquakes. Similarly, pear, apricot and peach trees have allegedly re-bloomed before earthquakes (Kamei, 2000). Just before the Great Kanto Earthquake in 1923 rice plants ready for harvest were shorter than usual. The leaf ends of some plants like rice have been said to become pale yellow or white before earthquakes and typhoons, producing striped horizontal yellow lines like a bar code. Before the 1976 Tangshan Earthquake in China, the unseasonal flowering and yellowing of bamboo, the food of the giant panda, led to some starvation of the species (Seismological Bureau of Anhui, 1978). Experiments on mimosa and rice plants are described in Chapter 6.

1.4 Unusual behavior of inanimate objects

1.4.1 Mysterious candle flame

The following old and familiar Japanese saying is mysterious and could easily be considered a superstition.

When candle flames on a Buddhist shrine or altar bend like archery bows, there will be a big earthquake.

Candle flames were found to bend in a laboratory experiment (Ikeya and Masumoto, 1997) described in Chapter 8 and shown on the book's back cover. There are stories that a confectioner predicted earthquakes by observing the colors of steamed rice used in Japanese sweets (Rikitake, 1976). Six hours before the Niigata Earthquake rice being cooked reportedly refused to come to the boil. These are also discussed in Chapter 8.

1.4.2 Nails dropping from a magnet and compass: A magnetic anomaly?

The *Ansei Chronicle* tells an interesting tale of magnets and nails. Iron nails about 15 cm in length, hanging from a big natural magnet as an advertisement in a spectacle store, detached from the magnet two hours before the Ansei-Edo (Tokyo) Earthquake (M6.9) in 1855. The magnet regained its iron-attachment property after the earthquake. The incident led to the immediate construction of an earthquake prediction apparatus [See Figure 1.6 (a)]. The apparatus does not appear to have been useful for earthquake prediction. A photograph of its laboratory reproduction is shown in (b) and the phenomenon was reproduced in a laboratory experiment (Ikeya and Matsumoto, 1997), described in Chapter 9. Milne (1890) referred to unusual movements in magnetometers at Paris in 1822, which coincided with slight shocks in Switzerland and the south of France.

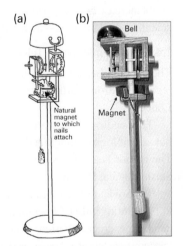

Figure 1.6 (a) A drawing in the *Ansei Chronicle* of the earthquake warning apparatus and (b) its reproduction in the laboratory (See also front color plates).

The north-pointing needle of a magnetic compass is also said to have moved violently before the Eastern Nankai Earthquake (M7.9) in 1944. A compass in a military avionic device on a Chinese Air Force MiG fighter jet malfunctioned when it flew over the imminent epicenter of the Tangshan Earthquake (Dai, 1996). The incident was explained as a probable earthquake-linked magnetic anomaly, but since the variation of the earth's magnetic field before earthquakes is known to be very small—one ten thousandth of the Earth's magnetic field (Rikitake, 1976)—these observations, if really observed, have to be attributed to some other cause.

1.4.3 Electric home appliances

Radio noise: Japanese listening to radio warnings of air raids by American B29 bombers in 1944 heard a lot of radio noise just before big aftershocks of the Eastern Nankai Earthquake (M7.9).

   Lamps, radios and clocks: A fluorescent lamp glowed faintly and radios did not work the day before the Tangshan Earthquake in 1976. Quartz clocks stopped nine hours beforehand (Dai, 1996).

   A variety of unusual malfunctions in electric home appliances were observed before recent earthquakes in Kobe, Izmit and Taiwan, as described in Chapter 2. The number of these reports in our increasingly electronically-driven world will no doubt increase and may soon produce a new crop of modern legends, unless we find scientific explanations—as we attempt to do in Chapter 9.

1.4.4 Modern legends: Telecommunication lines and apparatus

The evening before the Indian earthquake on December 15, 1872 Milne (1890) recorded a strong earth current on the telecommunication land lines from Bengal to London.

   The needles of electrometers swung to and fro before the Tangshan Earthquake but seismologists concluded the electrometer was broken and discarded the data. The author heard the story directly from a seismologist on a visit to the Anhui Seismological Bureau.

   The Soviet military detected electromagnetic noises before an earthquake in Iran in 1979 during military operations in Afghanistan, but took them for NATO electronic jamming exercises. Various noises at wide frequency ranges have been reported by telecommunication engineers and by those who study atmospheric lightning. The study of seismo-electromagnetic signals (SEMS) for earthquake prediction is a controversial issue but the evidence for it is growing, as discussed in Chapter 11.

1.5   Earthquake precursors and the Oriental worldview

1.5.1 Different ways of thinking in East and West

(a) The Oriental tradition

Generally speaking, the legendary earthquake precursor stories are not taken seriously in the West, though they may be related anecdotally in books on popular science. Most people in Japan—and presumably in China—accept the probability of unusual phenomena before earthquakes and simply think they are not yet understood scientifically. Asian traditional peoples, who place animals and even robots on the same level as human beings, or even consider them as messengers of the gods, generally accept legends about unusual animal behavior.

Japanese strongly influenced by Western culture still bear the stamp of their traditional culture. They are not deeply religious, but accept traditional religions and may customarily visit local shrines where e.g. a fox or white snake is worshipped. According to some Buddhist beliefs, human beings might be reincarnated in the form of some animal. Animism also imbues inanimate objects with souls. So it is not difficult for these Japanese to believe that animals have a heightened ability to pick up signals of an imminent earthquake and to interpret their behavior— particularly that of the catfish—as a warning.

(b) Scientists in Western culture: The "great ocean of undiscovered truth"

Western scientists, in the main, simply do not believe that animals can be used to forecast earthquakes. The old folk legends are treated as nothing but Oriental superstitions which must yield to an educated view of the world. This is an arrogant attitude because it is preemptive; insufficient scientific work has been done to establish whether or not animals are genuinely experiencing something science can measure. Unfortunately some westernized Asian scientists, influenced by their Western colleagues, have also been adopting the view that the legends are merely superstitions unworthy of serious scientific study.

Japanese scientists, who are half-westernized but still traditional in their attitudes are in a dilemma. They can appreciate that animals might have the ability to sense something before earthquakes, but their scientific training has made them skeptical.

Scientists should not close the door to further scientific investigation by convincing themselves in advance that it will be fruitless. We need to remember Sir Isaac Newton's modest dictum: beyond the shining pebbles and pretty shells that scientists, like children, have discovered, lies a great ocean of undiscovered truth.

1.5.2 Examining the Oriental approach

We often read articles on the gap in mutual understanding between East and West, Japan (or China) and the USA, and even Japan and China. We wonder if the proverb will always be the case: *East is East, West is West, and never the twain shall meet.*

But the gap between East and West is closing all the time. Our cultures are drawing closer. The West, in developing its pharmacological drugs from plant and animal extracts, is only doing what Oriental practitioners have been doing for thousands of years: producing herbal remedies from the leaves, roots and stems of plants and organs of some animals.

Earthquakes and Animals is an attempt to reduce the gap a little further between East and West, and the scientific and lay worlds, by subjecting the ancient Oriental legends and modern reports to scientific examination.

When new drugs are being tested, the drug under trial is given to one test group and a harmless substance with no therapeutic effect at all (a placebo) is given to another test group (called a control group). This is called a double-blind trial: neither patient, nor doctor, nor technician, knows which group has received the drug and which the placebo. Later when the results are decoded the true effect of the drug is known. What also becomes clear is that people who receive the placebo are also cured. This is called the placebo effect and it is significant in its effect. In other words—in the absence of other factors—people are capable of believing themselves into a improved state of health.

We need to test for the placebo effect in research into earthquake phenomena. If the scientific method reveals the earthquake legends are only legends then we have to accept that. But we may also find a real effect. The point is, someone has to put these legends (See Figure 1.7) to the test scientifically and see what remains at the end of the process. In the author's view what remains is the clear effect of electromagnetic signals, in effect "seismo-electromagnetic signals" (SEMS), on animals, plants and objects—the subject of this book.

Figure 1.7 Might the Oriental earthquake precursor legends so abundant in folklore and literary tradition, have a genuine basis in science? (A. Hochi.)

1.6 Summary

This chapter has introduced the reader to some of the many Japanese proverbs and legends about earthquake precursors. Whether they are real or merely superstitions is a matter of controversy. Most Asian people believe that animals are sensing something before earthquakes but that science cannot yet explain what it is. The author's view is that science is mature enough to explain what it is, but that specialists are too busy in, and sometimes protective of, their own fields to give themselves to the sort of joint effort that will clarify the phenomena.

Drawing from laboratory experiments *Earthquakes and Animals* attempts to demonstrate a scientific basis to many of the earthquake legends and reported precursor phenomena. The author is hopeful that animal precursors in particular will find a place in earthquake forecasting in an informal way. This is not to say there *will* be an earthquake, but the average citizen, aware of earthquake precursors, is more likely to take the kind of sensible precautions that will save lives and protect property.

2

Precursors Before
Recent Earthquakes
Kobe, Izmit, Taiwan and India

This catfish was photographed moving violently in its tank 20 hours before the Geiyo Earthquake (M6.7) on March 25, 2001, about 240 km from the epicenter. The normal behavior of a catfish is a motionless, resting state.

2.1 Introduction

After the 1995 Kobe Earthquake people in the area were asked to report any unusual phenomena they had observed before the quake. Their reports, gathered by mass media, letter, fax, phone and Internet, were published as a book, *1519 Statements: Precursors of the Kobe Earthquake* (Wadatsumi, 1995), which contained illustrations and photographs of earthquake lightning, vapor trails, elongated and red moons and unusual animal and plant behavior. The book was dismissed by some scientists as meaningless and misleading, but soon after the quake the Kansai Science Forum set up an interdisciplinary sub-forum whose brief was to study information on earthquake precursors with a view to saving lives. Its 40 members included physicists, seismologists, geologists, zoologists, telecommunications and information scientists and industrial sociologists.

Unusual phenomena were observed before the Izmit Earthquake in Turkey (Ulusoy and Ikeya 2001) and the Taiwan-921 Earthquake in 1999 (Ikeya *et al.*, 2000c). Questionnaires seeking more information were distributed in refugee centers and villages at the epicenters one month after the earthquake in Turkey and three months afterwards in Taiwan. Newspaper articles reporting unusual phenomena were also collected.

At the author's request, Dr H. K. Kundu (Geological Survey of India, Calcutta), investigated reports of unusual phenomena before the Gujarat Earthquake in India in 2001, and BBC listeners in Britain and the USA personally reported to the author their experience of unusual pre-earthquake phenomena after he appeared in the program: *Tomorrow's World*. Most of the accounts closely resembled the ancient legends and reports collected after the Kobe quake.

As already mentioned, in our technological age a new phenomenon is being reported: the unusual behavior of home-electric appliances before earthquakes. Clocks either stopped or their hands rapidly rotated, in either direction; radios, TVs, and air-conditioners suddenly switched themselves on, mobile phones rang but left no caller details. Naturally enough these odd stories invited comment ranging from the hilarious to the cynical.

Rikitake gave the name *macroanomalies* to unusual phenomena before earthquakes and collected stories about them from Japanese literature. These stories were indexed for epicentral distances and precursory times against earthquakes in China, San Francisco and other parts of the USA, countries of the former Soviet Union, Rumania, former Yugoslavia, Turkey, Italy and Central and South America (Rikitake, 1976, 2001). The database is helpful to scientists attempting to set up experiments or theories to test the possibility of a scientific basis to precursor stories.

The following reports from the Kobe, Izmit, Taiwan and Indian earthquakes are hopefully useful additions to Rikitake's database. Following sociological practice they have not been edited, because editing usually raises criticisms about the criteria for exclusion. They are the raw material, essentially as gathered from observers and survivors, and will appear naive and ill-founded in many cases. However, the reader is asked to look for common threads. It is these threads that are the subject, in *Earthquakes and Animals*, of scientific experiments and a hypothesis developed around an electromagnetic (EM) model of a geological fault.

2.2 What happened before the Kobe Earthquake?

2.2.1 Over 2000 statements on earthquake precursors

The Kobe Earthquake (M7.3) that occurred at 5:47 a.m., January 17, 1995, was caused by the movement of the Nojima fault on Awaji Island. Total official casualties were 6433, and over 40,000 were injured. The 1519 statements (1711 cases) on precursors collected by Wadatsumi (1995) through the mass media were mostly from the surrounding areas and can be classified thus:

Unusual animal behavior	872 (51%)
Sky and atmosphere	490 (29%)
Sea and land phenomena	189 (11%)
Electric appliances	149 (9%)

The Kansai Science Forum collected 173 statements from the epicenter (which Wadatsumi was unable to collect in the immediate post-quake confusion), but they were essentially the same as those he had already collected and similar to those described in Japanese legends and proverbs. As mentioned, the unusual behavior of home-electric appliances was a new feature.

2.2.2 Phenomena in the sky and atmosphere

(a) Earthquake light (EQL: 8 % of sky and atmosphere reports)

Figure 2.1 shows a photograph of the western sky in the morning under EQL. The sky again became dark almost immediately afterwards and the earth trembled. The phenomenon was observed almost coseismically by many people including graduate students and two colleagues, both professors. Other similar effects were seen before and after the quakes though it is difficult to distinguish these from lightning associated with thunderstorms.

(b) Earthquake clouds (EQCs) and earthquake fogs (EQFs) (44%)

Vapor trails (contrails): Photographers who photographed these at the time of earthquakes often claimed these were earthquake clouds (See Figure 2.1).

Tornado-like clouds: Some were photographed eight days before the earthquake (See Figure 2.2). Such clouds, some photographed half a day before the earthquake, and shown in the color plates, are discussed in Chapter 7.

Fogs: In spite of cloudless fine weather in the Kobe area before the earthquake, fogs were nevertheless observed in the region around Nishinomiya City, east of Kobe preceding the quake (Tsukuda, 1997). There were also reports of local rain from lay citizens, which puzzled a meteorologist in the Kansai Science Forum because it was quite inconsistent with the local weather conditions at the time.

(c) Sun, stars and moon (25%)

Sun: The morning sun looked unusually yellow.

Moon: The moon looked elongated (See Figure 2.3. Clouds appear to the left and bottom of the moon).

Stars: Stars felt so near that they could be touched.

2.2.3 Unusual animal and plant behavior

The following percentages are Wadatsumi's .

(a) Mammals (324 reports; 38% of reports on animals)

The Japan Pet Care Association distributed questionnaires to 210 owners living in 68 shelters in Kobe and found that more than one third (15/38 = 39 %)

Figure 2.1 Earthquake light (EQL) shown in a photograph of the western sky before the Kobe Earthquake, (Wadatsumi,1995). Vapor trails are also seen in the photograph (Photo: N. Yokota near Kansai Airport).

Figure 2.2 A tornado cloud photographed eight days before the Kobe Earthquake (Wadatsumi, 1995). A similar one was observed a day before the quake (See front color plates). (Photo: Ms T. Sugie.)

Figure 2.3 A photograph of an elongated moon with background clouds, taken at 8 p.m. on Saturday, January 16, a day before the Kobe Earthquake (Wadatsumi, 1995). (Photo: Mr H. Yamamoto.)

of cats and one fourth of dogs (39/149 = 26 %) behaved in unusual ways at the epicenter (Sugihara *et al.* (ed.), 1998).

Dogs (113; 15 %): Many stories from Japan, China and Europe tell of dogs saving human lives by drawing people outside before a disaster. In one TV account a man said he owed his life to his dog sleeping beside him before the earthquake. Other dog behaviors a day before the quake were described as puzzled or protective—as if a stranger were nearby. Some howled like wolves. Others refused to be separated from their owners, either insisting on staying inside or trying to get the owner outside sometimes early in the morning. Some left home before the earthquake and only returned several days later, with other dogs. Some dogs near the epicenter barked continually up to 30 minutes before the earthquake, waking their owners. Some dogs hid in corners, one in a bookshelf before the aftershocks. Others suddenly began to scratch at the floor or to dig up soil.

A neighbor said that for a week before the quake her dog refused to take its usual route towards the future epicenter area on its regular walks. Dogs often seemed to want to snuggle with owners just before aftershocks. The same dogs were also anxious before thunderstorms.

Cats (81; 11 %): Some cats tried to get into bed with their owners, waking them up; some bit their owners. Forty-five minutes before the quake others meowed to be let out of the house. Four abandoned cats, which usually appeared for food, didn't come the day before the quake but turned up afterwards at the family's half-demolished home. One cat, a Russian blue, usually gentle in nature, was violent one hour before the quake.

Sea lions (1) Zoo-keepers reported odd behavior by sea lions at the Kobe Oji Zoo about 25 km away from the north edge of the stressed Nojima Fault. They cried, refused to eat, jumped about, swam in zigzags and fussed. In a BBC report, *Tomorrow's World*, their keepers made jokes about an imminent large quake. An experiment on these sea lions is described in Chapter 4.

Hippopotami in the zoo submerged themselves before the quake and refused to surface for three days, according to the zoo's director, Dr M. Gondou. Their nervous keepers finally drained the pool and found them alive.

Squirrels: Zoo squirrels were found dead in their burrows after the quake.

Rats (63; 25%): A local rat trap caught a total of seven rats four days before the quake; the usual quota was none or one. People reported increased scuffling from rats before the quake, then they seem to have disappeared.

Hamsters either bit each other or their owners; some kept quivering—presumably in fear.

Rabbits: A fat rabbit, which rarely hopped or moved, suddenly became active.

(b) Human (91; 12 % of reports on animals)

Fatigue and irritation: A local professor noted in his diary a day before the quake that he had felt uncommonly tired and irritable that day.

Sickness, nausea, dizziness and sense of imbalance: Some people near the epicenter area reported symptoms such as dizziness, vomiting, motion sickness, hyperventilation, headache, fatigue and nervous irritability a day before the quake.

Strange odor: There were reports of a strange smell about 35 km away from the epicenter.

Cold touches: There were reports of a sensation like a cold touch on the cheek, tree leaves rustling when there was no air movement, and low frequency sounds just before the earthquake. The author noticed an ozone smell—always produced by high voltages—just before the quake.

Early waking: Many people in the earthquake zone said they had woken at about 5 a.m. that day—about 45 minutes before the earthquake—something they did not normally do. This is analyzed in Chapter 5, Section 5.8.

(c) Birds (281; 35 %) at the epicenter

Crows (102; 13%) flocked to a bamboo cluster at midnight, cawed loudly and restlessly (in keeping with the old proverb of noisy crows inviting disaster), and moved out to the suburbs (away from the epicenter).

Cocks crowed from 2 a.m. on January 17 in Okayama, 100 km from the epicenter.

Eggs: All eggs in a clutch laid before the earthquake had two yolks.

Seagulls disappeared from the epicenter and flew inland two days before the quake.

Sparrows (33 reports) disappeared from the epicenter before the earthquake.

Pigeons were seen rising in startled flight at 1 p.m., a day before.

Parrots normally talking frequently, fell silent and some showed signs of panic.

Pheasants screamed at 2 a.m. on Jan. 17 and again a few minutes before the quake.

Peacocks cried continuously for two days before the quake.

Parakeets flew round a cage in panic and one attacked another the day before the quake, but behaved normally again after the quake.

(d) Reptiles and insects (43 and 40 reports; 6 and 5 % of reports on animals)

Crocodiles clawed violently at the glass walls of their enclosure in the Kobe-Oji Zoo on the night of January 16, leaving scratches on the glass.

Snakes: A snake came out of its hole at a site 30 km away from the epicenter—an unusual event in winter.

Turtles woke prematurely from hibernation and attempted to climb the wall of an aquarium.

Stag beetles sleeping in wood chips woke up and emerged.

(e) Fish (93; 5 %)

A day before the earthquake, innumerable fish in a pond in Nishinomiya city, floated motionless near the surface, oriented east-west. Some fish were floating high or captured in great numbers, and others did not appear at all at various angling spots at the nearby beach. Some fish normally inhabiting surface water at the Osaka Marine Aquarium sank to the bottom and stayed motionless.

Deep-sea fish: Regalecus—popularly called the earthquake fish—normally frequenting deep water, was captured near the surface (See Figure 2.4).

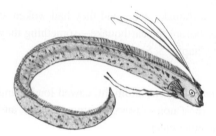

Figure 2.4 A drawing of the deep-sea fish, *Regalecus,* legendarily called the earthquake fish, a messenger from the dragon king's palace. This fish, up to 4–5 meters in length, is known to come up to the surface before earthquakes.

Catfish, loaches, flounder leaped violently out of their aquaria up to 2-3 days before the quake and again at 0:30 a.m. on January 17.

Dolphins at the Suma Aquarium, near the epicenter, moved nervously and leaped onto land a day before the earthquake.

Octopus: Many crammed into one trap. Some were found on shore moving with a staggering gait.

Squid: Local fishermen prophesied there would be an earthquake because of the large catch of squid in the month of December.

Fresh-water fish: Fish kept in a flowing water channel leapt out of water at the Aquapia facility about 60 km away from the epicenter and close to the Arima-Takatsuki tectonic line adjacent to the Nojima Fault. Numbers of fish behaving this way increased to nine in the period January 1 to 17, 1995, compared with two or three per month after the facility's opening in July, 1994. Species living near the bottom also leaped out, which puzzled the keepers.

(f) Plants (11; 1 %) swung in still air and flowered early.

Orchid: There were reports of a subtle swinging of orchid flowers and fluttering of leaves in windless conditions before the quake, and also of early flowering and re-flowering of some plants.

2.2.4 Unusual behavior of electric appliances—the "Alice in Wonderland" syndrome?

One new but characteristic feature was reporting of malfunctioning domestic electric appliances before the quake, presumably from electromagnetic interference. This was also evidenced in increased complaints from TV viewers: Japanese are demanding viewers, frequently registering complaints about noise on their TV screens from natural transmission disturbances caused by the appearance of sporadic E layers in the ionosphere in July.)

(a) Spontaneous on-and-off switching of electric appliances, and mysterious sounds

Fluorescent lamps: There were reports of lamps lighting dimly before the Kobe quake (as during earthquakes) and of mysterious buzzer and alarm sounds from intercoms and cellular phones.

Spontaneous switching on: Many people were surprised and puzzled when radios, TV's, and one cooler (though this was winter) turned themselves on from a few hours to a few days before the earthquake, sometimes around midnight or 2 a.m.

Strange sounds: Odd noises were reported from refrigerators and yogurt spoiled (possibly from spontaneous switching on and off of refrigerators).

(b) Noises on TV and radio and channel-setting anomalies

Some TV screens normally showing no speckle or flicker or electronic noise showed noise, "barber pole" color, distortion of the screen image, line noise and white bands before the earthquake (See front color plates and Chapter 9, Figure 9.3). Color screens turned black and white with image distortion; these were recorded on videotape. Remote controls worked erratically and TV sets fluctuated between channels, but worked normally again after the quake.

Truck drivers on highways near the epicenter could not pick up radio transmissions around 5 a.m., 45 minutes before the earthquake. (Professor Yoshino who first discovered electromagnetic (EM) signals before earthquakes (Gokberg *et al.*, 1982) traced this story.) Radio wave interference jammed broadcast radio waves along the fault line.

(c) Rapidly rotating clock hands

Quartz clocks showed fast or delayed times. One radio clock in Osaka (near Kobe) which normally readjusted automatically to GMT signals received by radio wave at 100 kHz every hour, was running two seconds slow before the earthquake, but functioned normally afterwards. Few will believe a similar and surprising report that the second hand of a quartz clock began to rotate quickly, first in one direction then in the other, a day before the earthquake.

Are these reported phenomena facts or invention? Can science explain them? Reproduction and simulation experiments on malfunctioning home electric appliances are described in Chapter 7.

2.3 Unusual phenomena before the Izmit Earthquake (M7.4) in Turkey

2.3.1 The Izmit Earthquake in 1999 and field survey

The Izmit Earthquake (M7.4) at 3:02 a.m., August 17, 1999, destroyed Izmit and Adapazari (See Figure 2.5), on the North Anatolian fault, one of the world's longest and most active strike-slip (horizontal motion) faults. The 1999 event is the 11th quake of M> 6.7 since records were first kept. Local soil conditions under buildings also affected the degree of shaking and ground failures. Dr U. Ulusoy (1999), who returned to Turkey after eight months in our laboratory as a post-doctoral fellow studying Electron Spin Resonance (ESR) of geological fault materials, asked citizens to report anything unusual they noticed before the earthquake. She collected 880 statements by 348 witnesses (male: 198, female: 150) by letter (105), fax (114), email (86) and phone (43). However, there were few reports from the epicenter area. So we visited the epicenter areas, Adapazari and Izmit one month after the earthquake and collected 137 statements directly from witnesses then living in tents and nearby villages. These proved to be very similar to reports from Kobe, and are

Figure 2.5 Land subsidence and destruction caused by the Izmit Earthquake.

classified in Figure 2.6. Scientists in Turkey seemed skeptical of the reports and critical of the publication, *Earthquake Precursor Data and Scientific Interpretation* (Ulusoy and Ikeya, 2001).

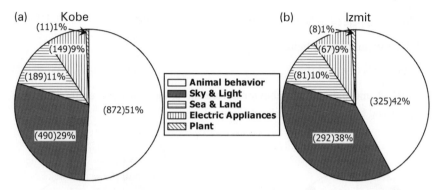

Figure 2.6 Comparison of precursor phenomena from (a) the Kobe Earthquake (Wadatsumi, 1995) and (b) the Izmit Earthquake (Ulusoy and Ikeya 2004).

2.3.2 Unusual atmosphere and sky

(a) Earthquake light (EQL) (208 reports:)

Many reported seeing light at the time of the earthquake. The color of the light was white and blue; sometimes reported as green at Izmit and red at Adapazari, 40 km away from the epicenter. Japanese NHK TV broadcast a documentary on the Turkish earthquake in which a scene at a gas station showed EQL at the epicenter. Some coseismic lightning in a big city might come from shorting of electric power lines during the shaking, but many reported seeing the light before the arrival of seismic S-waves or surface L-waves (produced by torsional oscillation).

For several days after the quake balloons of bright light came out of the sea over the Gulf of Izmit and the northeastern Marmara Sea and sounds of explosions were heard from the gulf area (Barka, 1999). Fire balls (presumably ball lightning) were observed several times during a period of two or three months before the earthquake, according to a fisherman (Ulusoy and Ikeya, 2002). Some fishermen described a co-seismic undersea explosion and light ascending out of the water into the sky. Fishing nets were found burned.

(b) Clouds and fogs

There were reports of striped clouds (contrails), black-gray fog and reddish-pink or orange skies. White fog was covering a graveyard when people left buildings at the time of the quake.

(c) Stars and moon

Stars: seemed unusually bright and extraordinarily close to the earth before the earthquake and also before the big aftershocks. Some seemed to move like comets with tails.

Moon: appeared to be reddish and vertically elongated.

2.3.3 Unusual animal behavior

(a) Mammals (291; 23 %)

Reports of unusual animal behavior before the Kobe and Izmit earthquakes are summarized for different species in Figure 2.7.

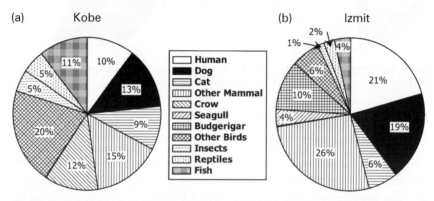

Figure 2.7 Animals showing unusual behavior before (a) the Kobe Earthquake and (b) the Izmit Earthquake (Ikeya and Ulusoy, 2000).

Dogs and wolves (106): Wolves (*Canis lupus*) howled mournfully for one and half hours before the Izmit Earthquake on August 17. They barked briefly 15 minutes before the M5.8 aftershock at 3 p.m. on September 13, the day the author arrived there. Dogs howled insufferably, as if they had rabies, or mournfully like wolves, barked loudly, or whined anxiously or panicked. Some barked and pulled their owners out of the house, others refused to eat, sought human company, ran off and then returned, or ran to the cover of trees in the park. They aggressively attacked other dogs, bit their owners, disappeared, formed packs and acted as if they were hunting an unseen enemy. They dug furiously at the ground, cocked their ears to the ground, looked up at the sky, and stood on their hind legs.

Cats (36) left home or disappeared for days before the earthquake. Some became restless, meowed unhappily, took kittens outside, climbed a high tree or wooden furniture, lay on their stomachs, or, legs splayed, moved in a cringing fashion;

climbed their owner's backs or bodies, insisted on protective holding by owners, panicked, bit their owners, and refused food.

Horses snorted, left their owner's farms, stamped the ground and jumped skittishly around.

Cows bellowed, crowded together, aligned themselves in one direction, refused to return to farms from mountain areas, were aggressive towards owners during milking, or seemed unusually happy to be near their owners.

Sheep seemed troubled as they grazed, bleating uneasily and crowding together.

Bats flew restlessly in circles.

(b) Human beings (116; 25 %)

People reported lost appetite, exhaustion, nausea, diarrhea, dizziness, nervous disorders, hysteria, facial ticks, high blood pressure, bleeding noses, headaches, cold feelings, pains in knees, bad weather symptoms, heart complaints and frightening dreams. There were reports of sensations of electric current in the brain or from finger to elbow, of an unusual amount of yawning and of disturbed menstruation.

A pregnant mother reported the movement of her baby in the womb suddenly stopped a few minutes before the earthquake.

Children woke and cried: The manager of a Toyota Car franchise in Adapazari said his son, a four year-old boy, woke his parents frightened that something was about to happen. His wife remarked that she had heard dogs howling like wolves. Five minutes later, the earthquake occurred.

Another Adapazari woman said her daughter of one and half years woke her up 15 minutes before the quake. She changed her clothing and both were awake and able to flee their home when the earthquake struck. A spastic child cried before the quake and laughed afterwards.

Most of the reports echoed the Japanese proverbs and stories collected after the Kobe Earthquake. Turkish interviewees were not aware of human symptoms before earthquakes. They were not aware of the Kobe reports before or during the survey.

(c) Birds (115; 25 %)

There were reports of unusual behavior on the day of the earthquake and the day before.

Budgerigars and parakeets (59) twittered in high tones, fluttered and woke up at 2 a.m., one hour before the earthquake. They kept to metal parts of the cage (contact with metal minimizes uncomfortable voltages) and did not go near their wooden perches. They shrieked, flew or walked at night, refused to eat or sing. For some reason such budgerigar reports were abundant in Turkey.

Tropical yellow parrots: These birds, similar to budgerigars, shrieked and panicked at the Darica-Kocaeli Bosphorus Zoo. Some experiments on these parrots are described in Chapter 4.

Seagulls (20) flew inland, flew restlessly about the sky crying, and flocked to roofs of buildings.

Crows stopped cawing, or gave weird calls, dived to windows and metal parts of cars and flocked to car roofs, although these were very hot.

Swallows twittered eerily, chirped "as if to warn their mates", flew restlessly round in circles, dived to walls of buildings but only on an east-west axis and attempted to find footholds on the wall.

Storks flew round in circles, then restlessly and prematurely migrated.

Cocks crowed at midnight, flapped their wings, panicked and shrieked.

Geese panicked, gobbled—as if in terror—and flapped their wings.

(d) Reptiles (13; 2.8 %)

Crocodiles at the Darica-Kocaeli Bosphorus Zoo (See Figure 2.8) refused to enter their pool.

Figure 2.8 These crocodiles at the Darica-Kocaeli Bosphorus Zoo, refused to enter their pool after the Izmit Earthquake. Though keepers were not observing their behavior before the earthquake it is possible the animals experienced the pool as an uncomfortable environment before the earthquake and were reluctant to re-enter it afterwards (See Chapter 4).

Lizards entered a house in unusual numbers.

Snakes swarmed in a garden.

Frogs stopped croaking and clung to the outside upstairs window of a house.

(e) Fish in the Gulf of Izmit (20; 4 %)

Fish and crabs: Starting two days before the quake, hundreds of fish, crabs and other living animals died. Visible fish numbers increased and some were seen in a state of panic. Fish lost their normal fear of humans, and floated vertically with their mouths open.

Deep-sea fish were seen swimming at the surface a day before the earthquake.

Crabs: Large numbers of crabs were found dead two days before the earthquake (Barka, 1999). Many crabs left their wet habitats, crawled ashore and were found in large numbers at a house 60 m inland. Some swam in the sea rather than remaining on the seafloor; the same behavior is described in Japanese literature.

Freshwater gastropods appeared at the seaside.

Jellyfish appeared two days before the earthquake (Barka, 1999) and also afterwards.

(f) Insects and worms

Ants: Left their nests, climbed trees and entered houses.

Bees aggressively stung humans and buzzed people to an unusual degree.

Cockroaches clung to upstairs window frames and hid close to metal ware.

Earthworms came out of the soil and even climbed to an upstairs floor.

Flies disappeared, bit aggressively, clung to human bodies and rotated as they did so.

Cicadas stopped chirping before the earthquake.

Mosquitoes either disappeared or greatly increased in number after the quake.

(g) Plants (23; 3%)

Plants dried up and leaves wrinkled. Plants grew slowly.

Reported animal and plant behaviors before the Izmit Earthquake are similar to those reported in Japan although most Turkish people were not aware of such phenomena. Some phenomena suggest clues to the nature of the physical disturbance sensed by animals and plants.

2.3.4 Unusual phenomena on land and sea (92; 10 % of total)

Land movement: People noticed higher sea levels, seaweed and mud deposits at the shoreline before the earthquake.

Wave sounds like the waves of the sea were heard underground near the seaside.

Death waves: Sudden and unexpected ocean waves splashed the shore areas. (The local people call these "death waves" because of their popular association with earthquakes.)

Monotonous weather (29): Very high temperature and humidity, with no rain, no wind; "boring" weather. People felt the effects of the sun more than usual.

Elevated water temperatures: Seawater, well water, hot spring water and ground water were unusually high in temperature.

Odor: People reported smells like sewage or burnt cables, both before and after the quake.

2.3.5 Malfunctioning electrical appliances (63; 7 %)

Reports of malfunctioning home electric appliances before the Izmit Earthquake comprised about 10% of total reports, similar to Kobe.

TV and radio (20): Radio noise and TV flicker and visual noise were reported as worse than usual before the quake.

Clocks (7): Three quartz clocks in different rooms, which normally showed identical times showed different times: fast, slow and normal before the earthquake. Afterwards people reset the clocks and they again synchronized perfectly. Another report claimed the second hand of a clock stopped.

Midnight Prayer: An unscheduled highly amplified Koranic prayer from a loud-speaker surprised people at about 2 a.m., one hour before the earthquake, possibly because a broadcasting device spontaneously switched on.

Phones and others (36): A man was woken at midnight by the ringing of his phone, but no one was on the line. The phone rang again and a few minutes later the earthquake occurred. He was unable to find anyone among his friends or relatives who had called him at that hour. Other reports about phones describe indicator lights on mobile phones lighting up and the phone ringing but no record of calls.

Powered car window: A car window opened and then changed its direction of motion. A refrigerator and washing machine more than 300 km from the epicenter made strange sounds.

2.3.6 "The Moses Phenomenon" (1; 0.1 %)

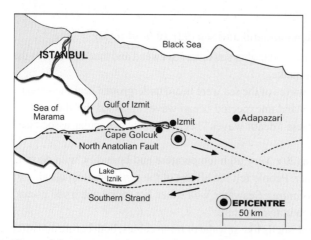

A strange incident was reported in which the sea parted near Cape Golcuk in the Gulf of Izmit and a boat touched the sea floor (See Figure 2.9). This was the only statement that differed from Japanese precursor reports. As the fisherman himself put it:

I heard strange sounds and felt vibration at 2.57 a.m. There was something strange

Figure 2.9 A map of the Gulf of Izmit, Turkey, indicating Cape Golcuk, the North Anatolian Fault and cities where the surveys were undertaken.

about the sounds of the screw and the vibration of the boat. I checked the screw, but there seemed to be no problem. Then I saw pinky-red color light streak from the sea surface into the sky. At about 3.04 a.m. the sea suddenly split in two and my boat went down and touched the sea floor. I saw walls of water on both sides of the boat, and then big waves bore me towards the Turkish Naval School buildings and I was carried to land.

The statement has interesting echoes of the Biblical account of the crossing of the Red Sea (Exodus 14:21) but could be explained by the local geography there as calculated theoretically and demonstrated in a model experiment in Chapter 8.

2.4 Precursors of the Taiwan-921 Earthquake (M7.7)

2.4.1 The Taiwan-921 Earthquake and field survey

The disastrous Taiwan-921 Earthquake (M7.7) struck the central part of Taiwan at 1:47 a.m. September 21, 1999, killing more than 2400 people. The epicenter was near the town of Chichi, Nantou County, along the Chehlungpu fault. (The earthquake was later renamed at the request of the Chichi townspeople.) Large-scale landslides and solid-soil liquefactions occurred, severely disrupting transportation, power and communication lines (See Figure 2.10).

The author visited Taiwan to collect precursor reports and collect samples of fault gouge from the surface fault ruptures for ESR analysis. Dr D. T. Shih at the Geological Survey of Taiwan helped us collect reports recorded in this section (Ikeya et al., 2000). Wadatsumi visited Taiwan immediately after the quake and collected stories through mass media advertising. In the following, the num-

Figure 2.10 Outcrops of the Chehlungpu Fault that moved at the time of the Taiwan-921 Earthquake.

ber of cases and percentages in parentheses are from Wadatsumi's home page (http://pisco.ouss.ac.jp).

2.4.2 Unusual atmosphere (320; 11 %), water (229; 8 %) and land (227; 8%)

A certain well showed an unexpectedly high water level of 4.0 - 5.7 m before the earthquake, but decreased considerably afterwards. Water became turbid three weeks before the quake and clear again three days afterwards. The same pattern was observed before the M6.5 aftershock a week later and the M6.8 earthquake one month later.

There were reports of ground to sky lightning, and low frequency sounds from the ground, before the tremor, which created a wave half a meter high in a lake.

2.4.3 Unusual behavior of animals, plants and electric appliances

(a) Mammals (623; 23 %) and birds (142; 5 %) in Taiwan

Dog: A dog howled repeatedly like a wolf—a frightening omen in Taiwan. The owners could not quieten the dog, which broke its chain before the quake and escaped. Its quarters were destroyed in the quake.

Dog hiding in an iron pipe: A mother dog with her three pups crawled into an iron pipe before the Taiwan-921 and would not come out. People thought the dog considered the pipe a safe haven. (Cockroaches and crows also sought out metal before the Izmit Earthquake.)

Cat: A cat which had made its home in a garage at the north edge of the Chehlungpu fault, moved out. A few days later, the quake destroyed the garage.

Monkeys: A large number of monkeys came out of mountains two weeks before the quake, screamed and shook trees just before the quake, and returned to the mountains three hours afterwards.

Panicked chickens: The Japanese newspaper, *Nikkei,* on September 24, 1999, carried a story that one and half million chickens in a chicken farm in a suburb of Shanghai city, about 1000 km from the ocean epicenter, panicked and attempted to flee their cages before the Taiwan-921 Earthquake. Two hundred thousand chickens were injured; twenty thousand died. Is it possible that even though they were so far away they still picked up some earthquake precursory signal?

(b) Fish (165; 6 %), reptiles (178; 6 %), insects (433; 15 %) and others

Fish: Anglers caught many fish during the night of the earthquake until they suddenly stopped showing interest in the bait and began leaping from the water.

Carp in a pond died from knocking their heads against the pond barrier.

Deep-sea fish: The Ribbonfish (Jordan & Snyder, 1901), with large eyes, no teeth and a red colored back is rarely caught, but a fisherman caught one month before the quake. One of the same species was also caught before a large earthquake in 2002.

Snakes formed swarms and were easily captured and cooked by local people.

Ants: A day before the quake ants near the fault were observed carrying their eggs to another habitat.

Earthworms: A large number of earthworms appeared above ground all over Taiwan before the main shock. Another swarm of earthworms

Figure 2.11 Earthworms emerged on the surface in large numbers in Taiwan just before the main quake and again some days before a large aftershock, when this photo was taken. (Photo: Mr Chen, Kaoshung Elementary School).

surfaced on October 25, eight days before a big aftershock (M6.9) on November 2 (See color plates and Figure 2.11). Chapter 4 describes an experiment on the site.

(c) Plant anomalies (24; 1%): Burnt plants and peculiar plant movements.

Plants: Plants with burned roots were observed at the northern edge of the fault.

Brown-colored tips: Blades of grass were found to be brown at the tips close to the fissure.

Upward pointing leaves: The leaves on a tree in an artist's garden at the north edge of the Chehlungpu fault pointed upwards before the quake when normally they hung earthwards. A warm current of air streamed from the ground to the sky in the garden where usually the air movements were cool. Grasses in the garden were brown at the tips. These plant behaviors were reproduced in experiments as described in Chapter 6.

(d) Electric appliances (124; 4%):

A policeman noticed rapid movement of the hands of his clock before the earthquake (but still managed to be late for his date with his girlfriend, according to the report).

2.5   The West Indian (Gujarat) Earthquake (M7.7) in 2001

2.5.1   Field survey and newspaper reports: EQL and sounds

There were many reports in local newspapers of unusual animal behavior before this earthquake (M7.7) in the Kutchch region, Bhuj, Gujarat at 8:46 a.m. on January 26, 2001 (Kusala *et al.*, 2001), but little academic attention was paid to them.

Dr H. K. Kundu, Geological Survey of India, Calcutta, visited the area and collected firsthand most of the following reports of unusual phenomena.

EQL: Dr M. Thakkar, Lecturer in Geology, R. R. Lalan College, Bhuj, was staying on the fifth floor of an apartment house. At the time of the earthquake, he made his way to the ground with his wife and two year old daughter. Just as the earthquake ended, he noticed the eastern sky was pale yellow in color. The color gradually faded and the sky resumed its normal blue-ash color.

Mrs M. Joshi was outside her house at the time of the earthquake and felt nauseous as her house and those about it began to swing. She remembers looking at the eastern sky and noticing it looked yellowish and reddish-yellow for a few seconds.

Sounds: The Internet posted the account of a man who predicted the earthquake on January 24 after he heard sounds emanating from the earth: a humming echo followed by rumbling sounds like heavy vehicles or falling objects, then sounds like thunder.

2.5.2 Unusual behavior of animals and plants

Dogs in a village called Baniyari, 53 km from Bhuj, barked loudly through the night of the quake (January 25-26) until the sun rose. The barking alarmed many people, but they did not expect an earthquake. Street dogs that normally ate food offered by villagers would not eat it that night but hid it in small pits and holes. In a TV interview a photographer in Bhuj said his pet dog was restless and barking incessantly two or three days before the quake, its nose constantly to the ground as if it were sniffing out something unusual.

Dogs and cows: The old walled town of Bhuj, about 42 km west of the epicenter near Bandhdi, has become home to many stray dogs and cows. The night before the quake they left the town and only returned to it afterwards.

Cattle: In a village near Dholavira, about 100 km from Bhuj, villagers noticed their cattle became distressed some 10-15 minutes before the earthquake and ran around randomly making "frightful" noises.

Scorpion: Dr A.G. Makwane, Lecturer in Physics, Bahauddin College, Junagarh, Gujarat, was at his agricultural farm in Gondal (50km from Bhuj), three days before the earthquake. He noticed a scorpion coming out of the ground, a very unusual event in wintertime, and told many people in Gondal and his colleagues in the college that something unusual was going to happen.

Domestic farm animals (cows, buffaloes and horses) refused to eat for three days after the quake.

Elephants: In the Ahmedabad Zoo elephants lowered themselves to the ground just before the earthquake.

Crows: A local taxi driver in the old walled city in Bhuj noticed that the local crow population deserted the roofs of the village one week before the earthquake, but suddenly returned afterwards.

Plants: Dr Thakkar found dried plants along a 9 km line on both sides of the Seeber Fault when they visited the area from February 1-14 (Rajendran *et al.,* 2001).

Snakes: In a TV interview, a farmer reported snakes in winter hibernation on his land coming out of their holes one or two days before the earthquake.

2.6 International mail and overall comments

2.6.1 Pigeons, a parrot and cats in the USA: Loma Prieta and Northridge

Pigeons were said to be agitated before the Loma Prieta Earthquake in 1989, leaving their usual roosting areas. Horses at a ranch were unusually skittish. However, scientists considered most of these statements wishful thinking. But after Granada TV broadcast our research on catfish, a former research biologist and avian behavior consultant living in San Francisco sent the author an email describing unusual parrot behavior before the Northridge Earthquake in 1994.

In 1994 there was a M 6.8 quake in Northridge, California—approximately 400 miles south of San Francisco. At approximately 8 pm the night before the earthquake my African gray parrot stared, as if in a trance, at the ground and resisted going into her cage. She hung upside down from perches, staring at the ground. The earthquake occurred at 4 a.m., the next morning. It was not felt in San Francisco. My parrot continued her strange behavior for approximately four hours after the quake. I have since noticed the same behavior for small earthquakes in the San Francisco area, and the intensity of her behavior and its length of time prior to the earthquake depends on the magnitude.... Several months ago there was a large quake in the desert of Southern California. I did a survey of parrots' behaviors just prior to and during the quake. The only bird that did react before the quake was a wild-caught gray parrot. I have an account from a person in Istanbul, Turkey whose African gray parrot demonstrated the same behavior several hours before the first large quake there.. . I also believe that her behavior is directed only to future quakes on the San Andreas Fault, because we live only a few miles from it. She did not respond to the recent Southern California quake that was on another fault. Regards, Jane Hollander.

Hanging upside down from perches might be an attempt to orient towards an enemy or insects—or towards a signal coming from the ground. Some literature mentions that even well trained police dogs howled, refused to obey commands, and kept their noses close to the ground as if sniffing (Raleigh *et al.*, 1977) apparently supporting such a hypothesis.

Dr Neil Whitehead wrote that a friend in Los Angeles said that his pet dog went "crazy" a few tens of seconds before the Northridge Earthquake, running madly about his house in a way he had never seen it do before.

A Web report on unusual animal behavior in the US says that the number of missing dogs and cats increases significantly up to two weeks before an earthquake and claims that earthquakes can be predicted both by looking at newspaper advertisements of lost pets and checking stresses in rocks caused by lunar gravity. Statistically the latter is still controversial.

2.6.2 Symptoms of a British former POW

The author received the following letter after the BBC program *Tomorrow's World* described experiments on animal behavior and detection of electromagnetic waves during rock compression (see Chapter 5). The incident occurred 60 years ago, just before the Japanese Nankai Earthquake in 1944. His letter reads:

> *In 1944 I was working on the deck of a partially constructed tug in the Kawasaki shipyard in Kobe. Standing erect and holding a large spanner, I experienced a sudden and nauseating attack of vertigo. I looked at my workmate kneeling on the deck and then around me; everything appeared to be normal.... A bare minute later, the deck began to move beneath my feet, I looked at my pal, his face was green. Amid shouts of "Jishin" (earthquake), all the Japanese working on the Tug scrambled down the ladders to the ground, and we wisely followed. Everyone ran towards the dockyard gates, why? I'll never know, for the overhead electric cables clashed and splashed, the ships' hulls rolled from one side to the other and the ground rippled like water. We would no doubt have been safe, finding a clear space. A few months before I finished the manuscript, the recent Kobe Earthquake occurred... my deep, deep sympathy for all the inhabitants who suffered in that catastrophe. Good luck, with your work Professor, maybe some of us humans get early warnings of Terra Firma's TANTRUMS. Sayonara (Good bye). Arthur Lowe, Lancashire, England.*

2.6.3 Overall comments on reports of recent earthquakes: No cultural differences

Turkey is an Islamic country, but close to Christianized Europe and quite different from Japan and Taiwan, which are both influenced by Buddhism. India is a predominantly Hindu country. We expected statements and legends to reflect different religious beliefs. But they did not. The replies of Turkish people to our inquiries about unusual phenomena before the earthquake were very close to those from Japan and China—and most of the Turkish eyewitnesses knew nothing of earthquake precursors. They all described crocodiles, parakeets or budgerigars behaving in similar ways.

Turkish people were puzzled that their electric appliances had malfunctioned and asked us if this could be linked to the earthquake and if so why. Concerned about media exaggeration we replied only that we were collecting people's observations and planned to analyze them later. We gave no details.

Although we took the advice of an industrial psychologist, Professor Kinoshita (Kansai Science Forum), in framing the Turkish survey, the serious involvement of the psychological profession would be valuable in framing and interpreting surveys to eliminate cultural or superstitious bias.

But, so far, because reports are so similar from all earthquake centers, it appears many of the precursor phenomena described are authentic rare natural phenomena.

2.7 Numerical data: fish capture records

2.7.1 Old work on fishery records by Terada

Skeptics argue that it is impossible to set up an effective objective measure of "unusual" animal behavior but one objective result is given by Professor Torahiko Terada, who checked records of fish catches. Unusual increases of fish catches before the 1923 Izu earthquake swarms were correlated as shown in Figure 2.12. If the figure is enlarged, it can be seen that increased catches preceded the earthquakes by four to eight days. Responding to some signal ahead of earthquakes fish seem to have left their usual habitats and gathered in groups in other areas where they were caught.

2.7.2 Sea bream catches before the Kobe Earthquake

Similar retrospective data just before the Kobe Earthquake in 1995 (Nakajima, 1996) are given in Figure 2.13. A possible precursor time of 6-8 days cannot be distinguished because of a break in record keeping over the New Year holidays, but the increase both before and after the earthquake is clear, as seen in (a). It is plausible that fish moved from the epicenter on the northern part of Awaji Island to the southern part where they were captured, as seen in (b). The data clearly indicate an unusual increase in fish capture; no such increase showing up anywhere in the remaining 10 year record. Daily records of buying and selling of fish were unfortunately discarded and only monthly totals were kept at the Osaka Central Fish Market.

2.7.3 Fishery records at the Istanbul market

Because the epicenter of the Izmit Earthquake extended into the Gulf of Izmit we decided to get access to daily fish catch records at the Central Fish Market in

Fish Capture (number)

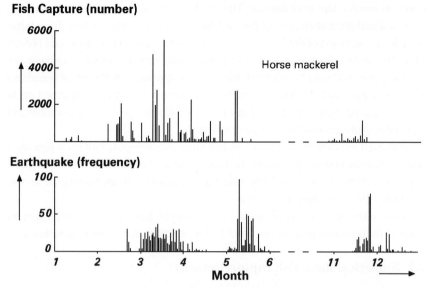

Figure 2.12 Fishing catches as a function of time before the Izu earthquake in 1923 and various seismic events (Terada, 1932). Although Terada made no note of it, the precursor times seem to be around a week.

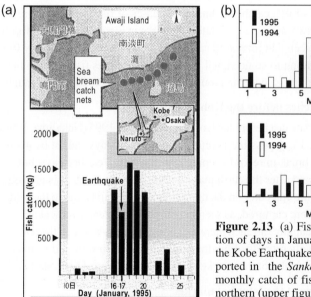

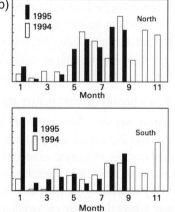

Figure 2.13 (a) Fishing catches as a function of days in January, 1995, in relation to the Kobe Earthquake (Nakajima, 1996, as reported in the *Sankei* newspaper). (b) The monthly catch of fish in 1994 and 1995 at northern (upper figure) and southern (lower figure) Awaji Island.

Istanbul—according to species. The daily variations in August, 1999 are shown in Figure 2.14 and compared with seismic events.

Unusually, a deep-sea fish, called Kofara, appeared at the market only three to four days before the 1999 Izmit Earthquake in Turkey. We are not sure how the fishing records should be correlated with the earthquakes but the data are capable of an interpretation consistent with Terada's big catches 4-8 days before the main shock.

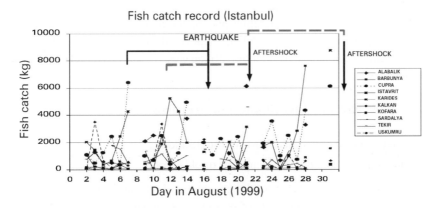

Figure 2.14 Fish catch records at the Istanbul Fish Market before the Izmit earthquake (M7.5), Turkey. The Izmit Earthquake occurred 9 days after the big catch. Following peaks might be either the precursors of large aftershocks or the consequence of microfractures of fault segments which also moved at the time of the mainshock (Ulusoy and Ikeya, 2001).

The Izmit Earthquake was the result of a few major ruptures triggered by fault movement. Each preseismic microfracture might have given signals leading fish to relocate and gather elsewhere—near a fishing ground—leading to a peak catch. Thus, the first peak catch may be eight days before the main shock and the two subsequent peaks may be associated with one of the big aftershocks. Some peaks might even be related to the later earthquake of November 2 (M7.2), though that earthquake was inland, while the Izmit Earthquake was close to the bay.

2.7.4 Comments on fish catch records

One remark should be made about precursory times. It could take a full day for fish to find their way to markets. The distance between the epicenter area and the catch site is also a factor. This makes the precursory times in Terada's work roughly concordant with times of other phenomena mentioned elsewhere in this book.

Sardines have also been captured in increased numbers before earthquakes, and deep-sea plankton have been found in their stomachs, indicating their movement close to the surface of the sea (Suehiro, 1976).

These studies suggest that fish catches might be useful in seismic prediction, but with qualifiers. Obviously fish capture depends on tidal flow, weather, and huge climate determinants like El Nino, etc.; larger than normal fish catches clearly cannot simply be attributed only to impending quakes. But if seasonal fish capture data for earlier years are studied for comparison, apparent increases may turn out to be identifiable earthquake precursors.

These difficulties in interpretation may be the reason why Terada's research was discontinued in Japan. But such fish catches could be monitored in real time, or records retrieved immediately after big earthquakes, for analysis for relationship to earthquakes.

2.8 Space and time distribution of precursors

2.8.1 Are unusual phenomena just wishful thinking?

We often encountered the following sort of objections.

There are hundreds of dogs in the epicenter and surrounding areas. Dogs bark all the time for all sorts of reasons—they might have fleas in their ears—and if they happen to bark the night before an earthquake that's nothing out of the ordinary. People feel tired and headachy all the time. Since there are millions of dogs and people, there are bound to be a number of such reports before earthquakes.

Or

Witnesses who have been alarmed or traumatized by an earthquake will connect everything with it. They will turn any trivial incident that happened before it into a precursor. If there had been no earthquake, they would never have remembered it. This tendency will decrease the further you get from the epicenter because the tremor also decreases.

On the face of it these seem reasonable objections, but there is sufficient time and location dependence of different kinds of phenomena that a correlation is clear. In other words the precursor reports are not completely random or invented. Statistical evaluation shows a significant correlation (See comments on chi-squared tests on the next page, and in Sections 5.8 and 7.6.2).

On dogs, specifically, veterinarians remarked that 20% to 30% of clients before the Kobe Earthquake who either called to seek advice about the odd behavior of their animals or brought animals in for care, said that they were behaving in unusual ways. This, compared with a normal zero level. The trend was sufficiently pronounced that two veterinarian co-workers, Drs M. Ohta and M. Hatoya, have collected blood from these dogs for genetic analysis to see if it might be possible to breed a species of "earthquake watchdog".

2.8.2 Spatial distribution of precursor phenomena

Animal species, distance from epicenter, and time before the main shock were plotted to see whether there were any correlations (Yamamoto *et al.*, 1999, Figure 2.15). A trend is visible: small birds appear to respond before larger mammals. A chi-squared statistical test on this figure shows the chance of these reports being sheerly random is vanishingly small (less than one in a trillion). The trend is therefore real; if it were only psychological the ratio of reports should have stayed constant.

Figure 2.16 shows phenomena reported at various distances from the Kobe epicenter. Reports of unusual bird activity increase just outside the epicenter area, then drop the further the distance from the epicenter, while reports of unusual sky phenomena increase according to distance from the epicenter. Possibly birds migrated away from the epicenter areas before the quake but the sky phenomena were mainly observable from long distances. There is more information on time and

Animals	Precursory Time (d)			Total
	$t < 1$ d	1 d $< t < 2$ d	$t > 2$ d	
Mammals	85	8	7	100
Birds	46	11	23	80
Total	131	19	30	180

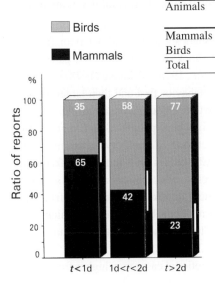

☐ Birds

■ Mammals

Figure 2.15 Unusual behavior of birds and mammals as a function of the precursory time (days) before the Kobe Earthquake (After K. Yamanoto in *Report of Survey & Research on Utilization of Earthquake Precursor Information* by the Kansai Science Forum, 1998). Error bars have been added to the figure.

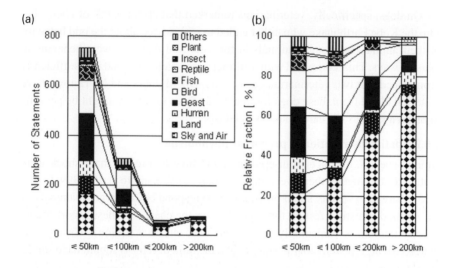

Figure 2.16 (a) Numbers of reports of unusual phenomena before the Kobe Earthquake, in relation to distance from the epicenter. (b) Their relative ratios normalized to 100. (Data taken from Wadatsumi, 1995.)

location dependence of precursor appearance of sun, moon and clouds in Chapter 7. For data on location-dependent and time-dependent human responses, see Chapter 5, Section 5.8.

2.8.3 A speculative mechanism for temporal shift of unusual behavior from small to large animals

The Chinese have observed that small animals show unusual behavior first and large animals later and closer to the main shock; this was observed particularly before the Haicheng Earthquake (Raleigh *et al.*, 1978). Rikitake (1978) noted that small animals like insects showed unusual behavior first and then larger animals e.g. rats, birds, cats, dogs, pigs and horses up to one hour before an earthquake. The animal size effect as shown in Figure 2.17 might be explained as an EM absorption effect [See Chapter 3, Figure 3.17 (b)]. Fracturing of small crystal grains in rock may lead to generation of sharp, seismic EM pulses with more high frequency components and a small pulse width, to which smaller animals are more sensitive. Only at a later stage when pressure is building up and larger rocks are fracturing will broader pulses with lower frequency components be generated to which larger animals and humans are more sensitive. [There is more on microfracturing and creation of EM waves in Chapter 3 (3.2.4, 3.3.7, 3.3.8).]

However this is speculation. Only in exceptional cases may higher frequencies

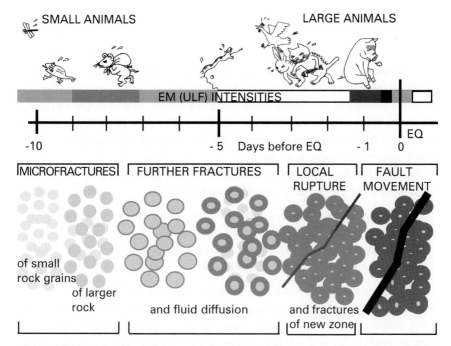

Figure 2.17 Showing the shift from smaller to larger animals over time, together with EM intensities (in gradients of black), and stages of rock fracture over the same period (with increasing fluid diffusion shown at the circle circumferences).

reach the surface (up a moist, rarely occurring conductive path). The much more likely scenario is that only ULF waves of similarly decreasing frequencies have a large enough skin depth (See Section 3.3.4.) to reach the surface. There they generate intense electric fields whose local discharges produce EM waves with high but decreasing frequencies, which are the waves directly responsible for the animal precursory behavior.

2.8.4 Times of precursor phenomena before earthquakes

The precise times of observation of precursor phenomena are hard to pin down since the reports are made retrospectively. However a theoretical formula has been proposed using Rikitake's empirical data for species-specific animal behavior for a certain magnitude (M) and distance from the epicenter (Ikeya *et al.*, 2001). Fraser-Smith *et al.*, (1990) proposed two peaks in precursor time for seismic EM waves at ultra low frequency (See Chapter 11, Section 11.3.1. and Figures 10.2 and 11.4).

Peak I: For frequent M7 earthquakes: from about two weeks before, peaking at 7 to 12 days before, and then decreasing.

Peak II: For M7 earthquakes: One or two hours before.

2.9 Summary

Reports of precursor phenomena gathered after four recent large earthquakes back up earthquake proverbs, legends and folklore abundant in the East, and these reports are tabulated in Tables 2.1 to 2.4 for the interest of readers. Whether these are genuine earthquake precursory phenomena or psychological in nature will always be controversial. However data showing time and location dependence of animal responses before earthquakes indicate that the reports have a more than merely psychological or superstitious basis.

In 1933 Terada wrote an essay entitled *A Group of Animals Regarded as a Group of Materials*. At the end of his article, he added this:

> *These studies are always regarded as heresy based on the conviction that physicists do not understand biology. What hinders the progress of science is not the ignorance of lay citizens but the lack of recognition of the essence and mission of science by scientists themselves. We must think deeply about this.*

Nothing has changed much in the last 70 years. One can talk about physicists instead of biologists, or biology instead of geophysics or seismology. Many physicists, geophysicists, biologists and seismologists are alike in their tendency to confine themselves to their sub-fields when science in the 21st century demands a much more unified approach.

Table 2.1 **Unusual animal behavior before earthquakes**

Mammals	Nervous, restless, irritated, panic and act "crazy".
Human	Headache, nausea, urge to vomit, giddiness, dizziness, heart complaints, nervous disorders, hysteria, bad weather symptoms.
Dogs	Bark loudly, whine ingratiatingly, anxious. Act as if tracking an unseen enemy; panic and bite owners, bark and pull owners outside, howl insufferably as if they have rabies.
Cats	Restless, meow pathetically, take kittens outside, climb high trees, jerk ears, lay ears back, leave home for days, disappear.
Rats	Disappear, fuss, panic, running on wires.
Horses	Stamp, snort, tremble, jump, buck, fall to the ground.
Cows	Bellow, crowd together, run away in panic.
Pigs	Aggressively bite each other, dig under fences, attempt to climb walls.
Deer	Leave bush and forest, do not fear humans, run to humans, run aimlessly.
Rabbits	Jump and run around.
Sea lions	Swim in zigzags, act agitatedly, fuss when out of the water, do not eat food.
Dolphins	Nervous, do not obey orders, leap out of the water.
Bats	Fly in the daytime.
Birds	Stop singing, become excited, flocks fly restlessly, cry weirdly, some die.
Chickens	Flap wings, shriek as if in terror, fly, fly to roofs. Cocks crow at midnight.
Hens	Lay no eggs, fewer eggs or eggs with two yolks.
Ducks	Avoid entering water, cry, act aggressively, bite humans.
Sparrows	Flutter in swarms, flutter down while flying, no twittering.
Seagulls	Fly inland, mew in sky, stay away from the sea.
Parakeets	High pitched chirping, flutter wings, fly at night, stay on fence, die.
Reptiles	Come out of hibernation.
Crocodiles	Call, leave the water for land or leave cages for the forest.
Snakes	Come out to the open in winter, swarm in bamboo clumps in summer.
Turtles	Wake from hibernation, climb on others apparently in panic, run.
Crabs	Leave wet habitats and crawl ashore, large numbers found dead.
Fish	Float and align in one direction, leap out of water, move violently, die. Turn upside down, act as if in turmoil, swarm, bigger fishing catches. Deep sea fish appear near surface, do not eat, sea fish swim up rivers.
Eels	Crowd onshore, disappear.
Insects	
Ants	Leave habitats carrying their eggs, swarm, enter houses.
Bees	Evacuate hives in a frenzy, buzz agitatedly and sting aggressively .
Cockroaches	Swarm close to metal ware.
Dragonflies	Swarm and fly in one direction .
Earthworms	Come out of soil, aggregate.
Flies	Swarm and cling to sweaty skin, fly in circles, rotate themselves.
Silkworms	Unusual alignment.

Table 2.2 Plant anomalies before earthquakes.

Blooming ahead of season		
Potato	Two months	Vines bloom.
Apricot	Six weeks	Trees bloom in winter.
Early flowering & early crops		
Rice	A few weeks before	Small plants, early crops, bar-code leaves.
Orchid	One day before	Sways without wind.
Mimosa	At or just before	Closes leaves and droops.
Tree leaves	Just before	Shake without wind.

Table 2.3 Malfunctioning home electric appliances before earthquakes.

Appliance	Behavior
Car navigators	Fluctuation of the pointer arrow.
Clocks (quartz)	Stopping or sudden movements of the second hand. Fast forward and backward movement or delayed movement.
Fluorescent lamps	Dimming of light as during thunderstorms.
Intercoms	Spontaneous buzzing sounds, or not functioning.
Mobile phones	Ringing & light but no record of caller. Do not function, make odd sounds.
Radio (AM)	Spontaneous switching and loud sounds, pulsed noise.
Refrigerators	Strange compressor noises, spoiled yogurt.
TVs	Spontaneous switching, speckling and flicker. Barber-pole color, lines, image distortion, white bands, loss of color, reversion to black and white, channel fluctuations.

Table 2.4 Earthquake precursor phenomena in the sky and atmosphere.

Phenomena in the sky	Preseismic (time)	Coseismic
Earthquake light (EQL)	A day or a few hours	Flash and arc just before
Earthquake fog (EQF)	A few hours or just before	Sudden dense fog
Earthquake cloud (EQC)	A few days, 8 days	A dragon cloud appears
Yellow sky	A day	Becomes dark
Short rainbows	A few days	...
Haloed sun	A day to a few hours	...
Elongated or red moon	A day	...
Stars appear close	A day	...

Note: Tables in this chapter may frequently be referred to, together with the proverbs in Chapter 1, if people wish to acquaint themselves with earthquake precursor phenomena.

Earth Sciences and Electromagnetism

A background knowledge of solid earth science and electromagnetism is necessary if we are to understand earthquake precursor phenomena. The complex interrelationship of the lithosphere, biosphere, atmosphere, and ionosphere, which bears on such phenomena, is diagrammatically reproduced above.

3.1 Introduction

An earthquake is a movement in the Earth's surface caused by a release of energy through a sudden dislocation of segments of the Earth's crust. There are a number of books on Geosciences dealing with earthquakes and plate tectonics, but it will be useful in this chapter to give a brief background in Solid Earth Sciences and Electromagnetics for the benefit of readers needing to understand the theory and experiments presented later in the book. Readers already familiar with intermediate level physics may skip this chapter.

The electromagnetic (EM) nature of some phenomena recorded before large earthquakes was discussed by Milne (1890) more than a century ago. In 1872, before the existence of earthquake detection equipment Milne reported that earth currents, strong enough to deflect a galvanometer needle, were observed in land telegraph lines before a quake. He also noted changes in the earth's magnetic field before earthquakes. Early on, the change was offered as an explanation for the Japanese report of iron nails falling from a natural magnet two hours before the Ansei Earthquake in 1855. In Chapter 9 this is ascribed rather to an electric field effect.

Lightning generated by atmospheric electric discharges produces EM waves. Recently EM waves have also been detected before earthquakes in frequencies ranging from ultra low frequency (ULF) to high frequency (HF). Seismo-electromagnetic signals (SEMS) have also been studied for use in short-term earthquake forecasting, but the results are controversial.

3.2 Solid Earth Sciences and earthquakes

3.2.1 Structure of the Earth and the cause of earthquakes

The Earth consists of three layers with distinctly different compositions. The core is composed mainly of metallic iron, solid at the centre and becoming more fluid from the center out. The core is surrounded by the mantle, a zone of dense, rocky but fluid matter, which is itself surrounded by the crust, which is of rigid rock but much less dense. The core and the mantle have a roughly constant thickness but the crust is far from uniform and differs in thickness from place to place. The crust beneath the ocean, *the oceanic crust*, has an average thickness of about 8 km, whereas the crust comprising the continents, *the continental crust*, ranges from 20 to 70 km in thickness with an average of about 40 km. The boundary between the mantle and the crust is called the *Mohorovicic discontinuity* or simply the *Moho plane*. It marks the base of the crust (Tarbuck and Lutgens, 1987; Skinner and Porter, 1987).

The thick continental crust is divided into upper and lower crusts, based on the different speeds of seismic waves through the two crusts. The upper crust is solid

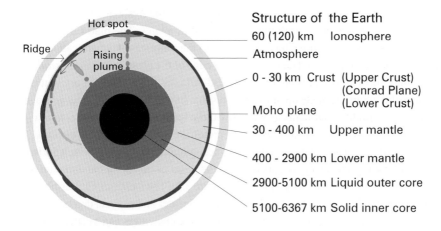

Figure 3.1 The layered structure of the Earth. Casualty-causing earthquakes occur in the upper crust.

and rigid, but the lower crust is ductile. The boundary between them is sometimes called the *Conrad plane*.

The atmosphere and ionosphere surround the Earth. The layered structure of the Earth is shown in Figure 3.1. EM waves, which are described later, pass through the less conductive upper crust and atmosphere and are also reflected by the ionosphere, ocean surface and the lower crust (Conrad plane), or the mantle (Moho plane).

The surface of the Earth is covered by a number of small crust fragments and six large rigid crust plates called *continental plates* which slide over the fluid-like upper mantle; the Eurasian, Pacific, North American, South American, Australian and

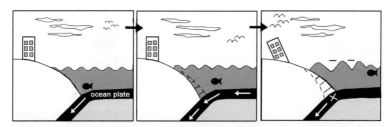

Figure 3.2 Subduction of an ocean plate under a continental plate. Rock fractures caused by seismic stress at the subduction zone of the plate boundary, and rebounding of the plate, cause large earthquakes, as in the last frame.

African plates. Some of the small crust fragments are the Arabian, Somali, Indian, Caribbean, Nazca, Cocos, Philippine and Antarctic plates. The plates move continuously and are sometimes forced down into the mantle at what are called *subduction zones* (Figure 3.2). The new growing edge of the plate, called the *spreading edge*, may form an ocean ridge (e.g. the Mid-Atlantic ridge). The theory that the surface of the Earth is made up of giant plates that have moved throughout geological time to form the continents as we know them today, is called *plate tectonics.*

Plate boundary earthquakes: These are earthquakes that occur in the upper crust at the boundaries of plates where seismic stress accumulates. The point at which the first release of earthquake energy occurs is called the *focus* or *hypocenter* and the surface area above the focus is called the *epicenter* (Figure 3.3). The location of an earthquake is described by the geographical position of the epicenter and its focal depth.

In a seismological map of the Earth, earthquake epicenters can be seen to form a long belt along the subduction zones of plate boundaries and these earthquakes are frequent. Most of them occur in areas bordering the Pacific Ocean called the *circum-Pacific zone*. This zone is also known as the *Ring of Fire* because volcanoes are often associated with the subducting edges of tectonic plates. About ten percent of the world's earthquakes are caused by the subduction of both the Pacific and Philippine plates under the Japanese Islands.

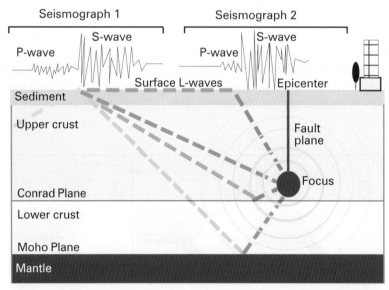

Figure 3.3 The upper crust, lower crust and mantle of the earth, indicating the focus (hypocenter), epicenter and fault. Seismographs at two different locations are shown with P- and S-waves.

Inland earthquake: A geological fault is a fracture in the earth's crust along which two blocks have slipped along each other several times in the past. Inland earthquakes occur as a result of sudden fault motions due to the accumulation of stress caused by relative movements of tectonic plates. The continental crust has innumerable geological faults.

Faults that move horizontally and vertically are called *strike-slip faults* and *dip-slip faults*, respectively. Very shallow earthquakes occur either when the internal stress overcomes the frictional resistance locking opposite sides of a fault plane or when brittle rock fractures in the crust.

The San Andreas Fault in California and the North Anatolian Fault in Turkey are well-known long faults causing large earthquakes. Ground movements are measured with the *Global Positioning System* (GPS)—also employed in car navigators—using radiowaves from satellites. The slip rate of about 15 mm per year at the earthquake prone northern strand of the North Anatolian fault suggests a 4 to 5 m displacement about every 300 years.

Magnitude (M): The energy of earthquakes increases 31.6 times per magnitude (M) increase of one unit and 1000 times per M increase of two units. In this book, we simply use the notation M7.3 for an earthquake with magnitude 7.3 (e.g. the Kobe earthquake). The number, frequency and magnitude of earthquakes are related in the Gutenberg-Richter equation (Gutenberg and Richter, 1956).

Earthquake volume: (the cubic measure of release of built up stress) is calculated by multiplying the stressed area at the epicenter and the stress drop. Briefly, a volume of 10 km x 10 km x 10 km corresponds to an earthquake of M7, like that of the Kobe earthquake, while laboratory rock compression of 15 cm x 15 cm x 30 cm in Chapter 5 corresponds to M-2. Earthquakes of M< 3, which can only be detected by sensitive seismographs, are occurring frequently everywhere.

3.2.2 Seismic P- and S-waves: Early warning systems and animal sensitivity

The energy of the stressed plate or block, is stored in a body of elastically deformed rock, then released by the movement of the geological fault. When the deforming force exceeds the limit of elasticity, the brittle rock in the crust fractures, leading to a major earthquake. A seismograph records the shocks and vibration caused by earthquakes.

Seismic waves travel through the earth's crust in two forms: compressional waves and shear waves. The compressional waves are also called *primary waves* or *P-waves*, the shear waves are called *secondary waves* or *S-waves*. The P-waves travel fast but cause little trembling; they pass through solids, liquids and gases. The S-waves cause the main shock but travel slowly and only through solids. They arrive at a seismograph after the P-waves. The velocities of seismic P- and S-waves are about 7 - 8 km/s and 3.5 - 4.0 km/s, respectively (we adopt 8 and 4 km/s for simplicity). Hence, one can determine the distance of the focus by multiplying 8 km/s by the S-P interval in seconds. Thus: if the S-P interval is 15 s, then the distance from the focus is 15 x 8 = 120 km. If animals are detecting P-waves, then the epicenter distance in km should be the time in seconds from the start of the unusual behavior, multiplied by 8. But our surveys show unusual animal behavior starting from a few hours to eight days before the arrival of S-waves, suggesting animals are detecting other signals.

In operational earthquake early-warning systems computers are set up to detect seismic P-waves so that precautions can be taken before the arrival of damaging S-waves. Other systems have been set up to detect P-waves: the Urgent Earthquake Detection & Alarm System (UrEDAS)—meaning, in Japanese, "beginning to tremble"— developed by railroad engineers and introduced to Japanese bullet trains (Shinkansen) in April 1992; a computerized early warning system called Caltech-USGS Broadcast of Earthquakes (CUBE), and the commercially available P-wave detection apparatus for early earthquake warning.

3.2.3 Foreshocks, main shock and aftershocks

It is impossible to know from seismograph data whether or not a small earthquake or earthquakes are foreshocks preceding a main shock. Earthquakes can only be described as foreshocks, main shock and aftershocks after the event. Although aftershocks have been said to have a different velocity ratio of P- and S-waves from that of main shocks, foreshock velocity ratio and seismic wave shape are not clearly different from those of main shocks in ordinary earthquakes.

Before the Kobe Earthquake small earthquakes occurred from north to south towards the epicenter, but it was only retrospectively obvious that they were foreshocks. A swarm of small earthquakes observed along the Ina River (where the author lives), before the Kobe Earthquake were thought by some seismologists to be foreshocks of the Kobe Earthquake. Others believed changing ground water levels in a small local dam over the 10 years since its construction might be responsible for the swarm.

3.2.4 The dilatancy-fluid diffusion model: Do microfractures precede earthquakes?

Dilatancy is an irreversible increase in rock volume caused by the formation of microcracks within rock. These cracks, which are oriented more or less parallel to the axis of maximum compression (Figure 3.4), open up, secondarily, in the direction of least compression (Kisslinger, 1974). The ratio of seismic velocities of P- and S-waves drastically changes prior to each earthquake due to dilatancy. The axial load at which dilatancy begins depends on the pressure and properties of the rock.

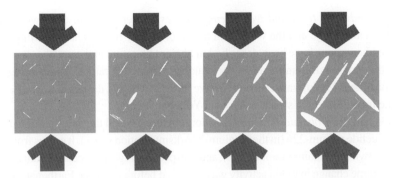

Figure 3.4 Open microcracks are formed in the rock and their openings grow as the stress increases (left to right). As these microcracks merge, macroscopic cracks are formed and then fractures. Movement along geological faults is stopped by large rock protrusions. Fluid diffusion may lubricate fractures and possibly generate electromagnetic phenomena, as described later.

As the stress in a seismically active region increases, the dilatancy may increase the effective pore volume in the rock. Initially this increase will cause the pore pressure to drop and "undersaturate" the rock fluid (cause the ratio of water to space to decrease). The ratio of P- and S-velocities changes as stress increases because S-wave velocity is much more sensitive to fluid saturation than P-wave velocity; S-waves cause a momentary change in the shape of the rock and cannot be transmitted through liquids and gas.

The dilatancy model was widely developed for earthquake prediction but the velocity ratio did not change as greatly as expected and seismologists have partly abandoned prediction of earthquakes based on it.

However, changes in local stresses around microcracks or fluid diffusion into cracks generate electric charges around faults, as described in Sections 3.3.7 and 3.3.8. It is possible that some precursory phenomena occurring ahead of earthquakes happen at the time of microcrack formation. If such phenomena are observed over a large area, then the earthquake volume will be large, indicating a large magnitude.

3.2.5 Geophysical and geochemical changes before earthquakes

Studies of earthquake prediction have been largely confined to the earth's geodynamic activity and the following observations have been made by geophysicists and geochemists.

Ground tilt: Before a volcanic eruption, a volcano swells measurably in response to upward movement of molten rock called *magma*. In other words, the ground tilts. Tiltmeters (devices that measure ground tilting) are also used in short-term earthquake forecasting based on tilt caused by plate rebound.

Groundwater level and water temperatures: Water levels in deep wells and springs can change and the water turn turbid or muddy before a large earthquake. These phenomena have been described in ancient literature and were observed before the Kobe, Izmit and Taiwan earthquakes. The temperature of hot springs at Hakone (near Mt Fujiyama) is said to have increased before the Great Kanto Earthquake in 1923 (which struck Tokyo and Yokohama).

Gas release from microfractures: Radon is a rare gas and the daughter of natural radionuclide radium-226 (in the uranium-238 disintegration series). It comes from the ground (which contains uranium-238), and is especially high at brecciated fault zones. An extreme change in radon activity was observed before the Kobe Earthquake (Figure 3.5). It rose to 20 times normal levels and suddenly decreased 7 days before the earthquake, returning to pre-October, 1994 levels afterwards (Silver and Wakita, 1996).

The variation of radium activity in mineral water bottled from a well near the epicenter of the Kobe Earthquake followed a similar pattern: an increase and return to normal levels (though the return to normal radium levels occurred after the earthquake (Igarashi *et al.*, 1995). The same tendency was observed before the Taiwan-921 Earthquake (Yang *et al.*, 2000). Possibly microfractures in rocks release radon and radium into ground water before fault movements, i.e. earthquakes.

Release of gaseous hydrogen and carbon dioxides have also been detected before earthquakes, and an anomalous increase of nitrogen oxides a week and a day before the Kobe Earthquake was recorded by the local Environmental Protection

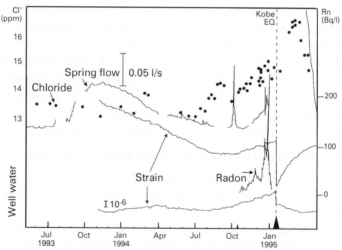

Figure 3. 5 Concentrations of radon and chloride in well water before the Kobe Earthquake (revised after Silver and Wakita, 1996).

Agency. However, air pollution in a large city makes it hard to call these elevated levels earthquake precursors; the high levels of CO_2 and NO_x measured after the Kobe Earthquake also coincided with post-earthquake traffic jams (Matsuda *et al.*, 2000). Links between gas geochemistry and earthquake precursors have been reviewed (Toutain and Baubron 1999; King, 1984).

Intrafault materials: The internal structures of fault zones, fault development and behavior have been studied for insights into the mechanics of earthquake rupture, fault zone evolution, and strong ground motions. Such a project, the Nojima fault-drilling scheme was initiated after the 1995 Kobe Earthquake. Geological, geochemical, and geophysical surface measurements have helped revise our view of fault zone structures. Intrafault material consisting of fault gouge (clay) has been studied using Electron Spin Resonance (ESR) (Ikeya, 1993).

3.2.6 Tidal withdrawal, "sea-splitting" and tsunamis

Water flow directions have been known to reverse in rivers before earthquakes. Preseismic tidal withdrawal has often been described in the literature. Water soaking into a dilatant area, and land-tilting might play a major role in such rapid tidal withdrawal.

The odd sea splitting effect ("The Moses Phenomenon") described at the Gulf of Izmit (See Section 2.3.6) is explained experimentally in Chapter 8 and might be related to a primary process of *tsunami* formation

Tsunamis are sea waves caused by earthquakes beneath the ocean floor. They sometimes travel at speeds of 1000 km/h (280 m/s) and can devastate coastal regions. The detailed mechanism of tsunamis is still under study.

Those wanting to learn more about geophysics and earthquakes should consult textbooks of general geology and publications issued by national Geological Surveys.

3.3 Elementary Electromagnetism

3.3.1 Physical quantities and units

(a) Charge and electric current

Electric charge occurs in nature in two forms, negative and positive. Two particles that have similar charges (both negative or both positive) repel each other, whereas those that have dissimilar charges (negative and positive) attract each other. The unit of a charge is the amount of charge on an electron, $-e$, which is equal but opposite in effect to the positive charge on the proton, $+e$. Charge is expressed in SI (Système Internationale) units as a *coulomb* [C]. A negative ion (an atom or group of atoms bearing more negative than positive electrical charges) has more electrons than the constituent neutral atom, while a positive ion has fewer electrons.

A flow of electric charge is called *current* and is expressed in SI units as *ampere* [A]. An ampere is equal to one coulomb/second [C/s]. One milliampere [mA] is one thousandth of 1 A. A current of 0.002 mA causes muscle contractions in animals such as frogs. (Galvani first observed muscle convulsions in frog legs on application of electric stimuli of at least this amount more than 200 years ago.) Hospital patients inserted with probes and needles are susceptible to 0.02 mA (Sternheim and Kane, 1986). The physical quantities of electromagnetism and their symbols and SI units are summarized in Table 3.1.

Table 3.1 The SI unit for the physical quantities in electromagnetism

Quantity	Symbol	SI Unit		Comment
Charge	Q	coulomb	[C]	One billion electrons have 0.16 nC
Current	I	ampere	[A]	Charge flow of 1 coulomb/second [C/s]
Potential	V	volt	[V]	Earth potential difference: 0.01 mV/m
Electric field	E	volt/meter	[V/m]	Sensitive animals: 0.01 mV/m
Magnetic field*	B	tesla	[T]	1 gauss = 0.1 mT: B_{Earth} = 0.035 mT
Frequency	f	hertz	[Hz]	Radio AM: 100 kHz FM: 100 MHz
Resistance	R	ohm	[Ω]	Ohm's Law: $I = V/R$
Impedance	Z	ohm	[Ω]	$I = V/Z$, Z depends on the frequency f.
Power	P	watt	[W]	Joule/second [J/s] $W = V \times I$
Power flow	S	watt/m²	[W/m²]	S of EM waves is Poynting's vector
Energy (Work)	W	joule	[J]	$J = W t$ 1 cal = 4.186 J

* Magnetic field is actually expressed by H using the SI unit of amperes per meter [A/m]. However, the magnetic flux density (symbol B) in units of nanotesla (nT)—one millionth of 1 mT—is used for the intensity change before earthquakes.

(b) Electric potential and electric field

Electric potential is the energy required to bring a unit of electric charge of one coulomb from infinity to any point in an electric field. It is expressed in SI units as *volts* [V]". The potential difference between the plus and minus terminals of a battery is typically 1.5 V. A line or plane formed by positions of similar potential is called the *equipotential line* or *equipotential plane*. A conductive plane like the ocean forms an equipotential plane. A fault zone with breccia might form a conductive plane because of dissolved minerals.

The electric potential gradient (the slope between two points with different potentials) is called an *electric field*, and is expressed in SI units as *volts per meter* [V/m]. Positive and negative charges produce the potential difference and are, respectively, a source and a sink of the electric fields. (See Figure 3.6 (a) for an electric dipole, or pair of separated charges.)

If the voltage of a dry cell, 1.5 V, is applied to two parallel plates separated by 15 cm (we use meters in SI units: 0.15 m), the electric field is obtained by dividing the voltage by the distance, i.e. 1.5 [V]/0.15 [m] = 10 [V/m].

(c) Magnetic field

A magnet produces magnetic field lines running from the north pole (N) to the south pole (S) outside the magnet [Figure 3.6 (b)]. It orients a compass, a permanently magnetized needle pivoted at its center. The strength of magnetic fields (actually magnetic flux density) using the symbol *B*, is expressed in SI units as *tesla* [T]. The tesla unit is too large compared with most magnetic fields in everyday life, and so the unit of one thousandth of a tesla, mT (milli-tesla: 0.001 T) is used in this book, the most frequently appearing unit being one bil-

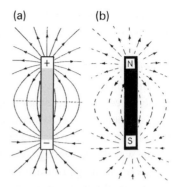

Figure 3.6 (a) Electric fields from a positive charge to a negative charge are called *electric dipoles*. An electret is a permanent electric dipole used for microphones. (b) Magnetic fields from the north pole (N) to the south pole (S) of a magnet are called *magnetic dipoles*.

lionth of a tesla (nano-tesla: nT), i.e. one millionth of 1 mT. The observed change in the magnetic field before earthquakes is generally a few nT, a few one-tenths of a thousandth of the Earth's magnetic field (0.05 mT), or about 100 nT before large earthquakes.

Magnetic fields are also produced by an electric current. Small electric currents associated with the human nervous system produce very weak magnetic fields: about 10 pT near the chest and 0.1 pT at the surface of a single nerve fiber or axon. [One billionth of 1 mT or one thousandth of 1 nT is called 1 pico-tesla (pT).] The *magneto-cardiogram* (making use of magnetic fields) is used as the equivalent of the better known *electrocardiogram* for diagnosis of heart and brain conditions.

(d) Resistance and impedance

According to Ohm's law, an electric current is proportional to the applied voltage and inversely proportional to the resistance, which is expressed in the SI unit as an *ohm* [Ω]". That is, an electric current of 1 A flows by the application of 1 V to a resistant material of 1 Ω. Normal dry skin has a resistance of 1 million ohm (1 MΩ) and detects a voltage of 10 V when a current of 0.01 mA flows in it. Sweaty people have less resistance, so are more conductive and sensitive to electric fields. Electrical current produced by one hundred volts in a commercial or domestic setting is life-threatening for a wet or sweaty person. Animals without electrosensory organs respond to a current of 0.01 mA, much like Galvani's frog muscles (See experiments described in Chapters 4 and 5).

The resistance to alternating current (AC) is called *impedance*. Impedance has direct current (DC) and AC components. Its magnitude is also expressed in ohms [Ω]. Impedance depends on the frequency and becomes very small at a particular frequency called the *resonant frequency*. Currents in animal bodies depend on the body's impedance (i.e. its size and conductivity) as well as on the frequency of the AC voltage. They also depend on the ambient environment (where the animals are) and the way the AC voltage (electric field) is applied.

Electrical energy is used only by the DC component (the resistance part) of the impedance. When two circuits are connected, the energy is transferred most efficiently at a particular frequency at which the total impedance is low, and this resonant situation is called *impedance matching*. Induced currents in animals affect the animals rather than the electric fields themselves and the hazardous effect of AC electric fields depends on the body's impedance level, which depends on the AC frequency and the size of the animal.

(e) Conductivity and EM anomalies:

Geophysicists measure electric potential changes and changes in electrical conductivity by applying a voltage. Before volcanic eruptions, presumably due to stress changes and fluid movement caused by upward movement of magma, EM anomalies have been observed similar to those reported before earthquakes. Seismo-electric magnetic signals (SEMS) will be discussed in Section 3.3.3 and in Chapter 11.

3.3.2 Electromagnetic (EM) waves

(a) Continuous wave (cw-) EM field

Both electric and magnetic fields are connected in electromagnetism as described by Maxwell's equations. These hold that when magnetic fields change, electric fields are generated. EM sine waves are called continuous waves (cw) and are propagated with the electric field and magnetic field perpendicular to each other (Figure 3.7).

(b) EM waves and their frequencies

Long wave EM waves are ultra low frequency (ULF), extremely low frequency (ELF),

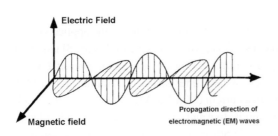

Figure 3.7 Electromagnetic waves having electric and magnetic field components.

very low frequency (VLF) and low frequency (LF). Middle frequency waves, that is, high frequency (HF), very high frequency (VHF) and ultra high frequency (UHF)] are used in radio and TV transmission. Super high frequency (SHF), millimeter waves and submillimeter waves are called *microwaves*. A microwave oven employs 2.45 GHz (1 GHz = 1000 MHz = 10^9 Hz), which vibrates water molecules efficiently to heat food. Mobile phones employ 800 MHz but have now expanded into the GHz range. Visible, ultraviolet and infrared light are a sort of higher frequency EM wave.

For the purposes of this book i.e. the discussions of unusual animal behavior in Chapters 4 and 5 and SEMS in earthquake prediction in Chapter 11, we deal only with EM waves at lower than microwave frequencies.

Figure 3.8 categorizes cwEM waves. They are named according to their different frequency ranges expressed in units of hertz [Hz] together with their usage.

Frequency	Name	Use	Wavelength
	Radio waves	Telecommunication	
1 Hz	ULF	Submarine	300,000 km
	ELF	Ship	
	VLF	Omega navigation	
100 kHz	LF	Decca navigation	3 km
1 MHz	MF	AM radio, Commun.	
10 MHz	HF	Short wave commun.	30 m
100 MHz	VHF	TV, FM radio	
1 GHz	UHF	Mobile phone	30 cm
	Microwave	Microwave oven	3 cm
	SHF	Satellite broadcast	
	mm-wave	S-commun. Radar	
1000 GHz	Sub-mm wave	Radar	0.3 mm
	Light: Far-infrared	Heater	
	Infrared light (IR)		770 nm
1,000,000 GHz	Visible light	Light commun.	
	Ultra violet	(UV) light Sterilization	380 nm
	X - rays	CT diagnosis	10 nm
	Gamma-rays	Food irradiation	10 pm

Figure 3.8 The name and frequency of continuous EM waves together with their uses in the modern world. The unit of nm (nanometer) and pm (picometer) are one billionth (10^{-9}) and one trillionth of (10^{-12}) of one meter.

The power of EM waves is expressed in W/m² or by their electric field intensity in V/m or their magnetic field intensity in nT. Magnetic storms induced by solar flares sometimes cause changes of 300 - 400 nT. Changes in the magnetic field before volcanic eruptions and large earthquakes are from 1 - 100 nT. Table 3.2 gives conversions from a magnetic field of EM waves at ultra low frequency (ULF), to power, and to electric field. The conversion from the magnetic field to the electric field is obtained by multiplying by $c(3 \times 10^8)$ $[E = cB]$. A magnetic field of 1 nT gives an electric field of 0.3 V/m.

Table 3.2 Electric and magnetic fields of ULF EM waves and their power flow

Electric field E [V/m]	Magnetic field B [nT]*	Power S [W/m²]
1.0	3.3	0.0025
0.3	**1.0**	0.00022
19.4	65.0	**1.0****

The conversion equation: $E = cB = 19.4\ S^{1/2}$, where c is 3×10^8 m/s and B in [T].
* 1 nT (nano Tesla) = 0.000,001 mT = 0.000,000,001 T.
**The level of power required to cause unusual animal behavior in ordinary animals and malfunctioning of some electric appliances.

The modern environment is full of EM waves, either created by man e.g. emitted by cellular phones and broadcasting, or naturally produced e.g. by lightning and before large earthquakes. In early 2000 a mobile phone generated EM waves of 0.8 W at 800 MHz [about 10 W/m² ($E = 60$ V/m; $B = 200$ nT) at a distance of 5 cm from the antenna]. At this level, some electronic appliances malfunction, some animals respond and some people are said to complain of headaches after long usage. However, for common EM waves, the power is small a long way from the source, as shown in Table 3.3. So only mobile phones might have hazardous effects.

Table 3.3 Typical power and power density of EM waves in our daily lives. The distance from the transmitter is assumed to be 10 km for radio and TV and 1 m for mobile phones.

Transmitter	Power P	Receiver $S = P/4\pi R^2$ (mW/m²)	Electric field $20S^{0.5}$ (mV/m)	Magnetic field $B = cE$ (nT)
Radio AM	10 kW	0.01	60	0.2
FM	1 kW	0.001	20	0.06
TV video	10 kW	0.01	60	0.2
audio	2.5 kW	0.002	30	0.1
Mobile phone	0.8 W	60	5000	20
PHS	0.02 W	1.5	800	3

(c) Unit of decibel [dB] and power [dBm] used by electrical engineers

Signal in [dB]: It must be noted that decibel intensity does not become zero at a point greater than the skin depth (for skin depth, see 3.3.4). If the power, which is proportional to the square of the electric field intensity, decreases by one order of magnitude, electrical engineers say it has a signal of -10 dB. If it decreases to one hundredth, one thousandth, one millionth, it is described as -20 dB, -30 dB and -60dB respectively. Thus, if power has decreased by one hundredth and is described as -20 dB, the electric field intensity will be one tenth. If an amplifier circuit amplifies the weak receiver signals one million times in energy (one thousand times in the electric field), the amplifier has a gain of + 60 dB. (Some circuits can amplify up to 120 dB!)

Power in [dBm] and [W/m²]: Absolute power is sometimes expressed in units of dBm based on 1 mW of transmitter power. Refer back to Table 3.2 for the power (energy flow) in W/m² for different intensities of electric and magnetic fields of EM waves.

3.3.3 Electromagnetic pulses and seismo EM signals (SEMS)

Changes in the electric dipole moment, i.e. the product of positive or negative charges and their separation distance, result in the generation of EM waves according to Maxwell's equations. For instance, a plasma, i.e. an ensemble of positive ions and negatively charged electrons, produces EM waves. We believe that charges generated by microfractures in rocks in earthquake zones produce EM pulses prior to the main shock, although the mechanism of charge generation is still controversial at this time. Such EM waves are an ensemble of pulses.

An EM pulse is a packet of EM waves composed of different frequencies and of short duration (Figure 3.9). A pulsed EM field is different from that of a cw-EM field in its effect on EM receivers, including animal sensory organs. When an EM pulse is received by a receiver it passes

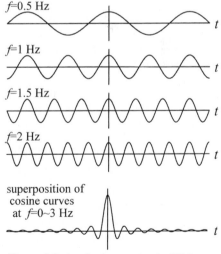

Figure 3.9 A pulse is composed of EM waves at different frequencies and expressed as their sum. Pulsed noise is called *chirps*.

to the detector a pulse known as "chirps" or "chirp waves" in a wide range of frequencies.

3.3.4 Skin depth of EM waves

The intensity of electric and magnetic fields exponentially decreases when the EM waves are propagated in a conductive medium. The point at which the intensity becomes 0.368 of the initial value is called the *skin depth*. The skin depth decreases as the frequency of EM waves increases. Unfortunately most textbooks on geophysics only give a skin depth equation for metal, which misleads many geophysicists. The earth's crust is not as conductive as a metal and an exact equation of skin depth must be used in discussing the propagation of EM waves. Typical skin depths calculated for geological materials using each resistivity (ohm-m) and a specific dielectric constant, $\varepsilon*$ (the amount radiation is absorbed or dispersed) at different frequencies, are given in Table 3.4. The calculated intensity of the EM waves at 10 times the skin depth is still not zero, but about six one hundred thousandths of the original value (6/100,000 = -43 dB), which can still be detected easily by modern detectors with amplifiers.

Table 3.4 Skin depth of EM waves at different frequencies for typical resistivity, ρ and specific dielectric constant, $\varepsilon*$ of rock, sediment, water and sea water.

Material	$\varepsilon*$	ρ [ohm-m]	$\varepsilon\rho$ [s]	Frequencies [Hz] and skin depth [m]					
				f= 0.1 Hz	1 Hz	10 Hz	1 kHz	1 MHz	1 GHz
Granite	8	10k	7×10^{-7}	159k	50.3k	16k	1.6k	151	15.0
	8	1000	7×10^{-8}	50.3k	15.9k	5.04k	504	19.8	15.1
Sediment	40	100	3×10^{-8}	15.9k	5.03k	1.59k	15.9	5.62	3.36
Water	80	50	3×10^{-8}	11.3k	3.56k	1.13k	113	3.97	2.37
Ocean	80	0.25	2×10^{-10}	800	252	79.6	7.96	0.252	0.013
Crust & soil*		10k	- - -	164k	65.6k	16.4k	1.60k	151	150
	$\varepsilon*$ (f)		$\varepsilon*$	1000k	1000k	10	8	8	8

* The frequency dependence of the dielectric constant for earth crust (dispersion) takes into account the skin depth. The symbol k in the table indicates x 1000.

3.3.5 Condenser and dielectric materials

When a battery applies a voltage to two metal-plate electrodes, charges appear on the electrodes as shown in Figure 3.10 (a). The plates and intervening space are called a condenser (or a capacitor). When a high voltage is applied to two slightly separated electrodes, the electric field intensity (in units of [V/m] obtained by dividing the voltage by the separation distance in meters) becomes large. If it exceeds 3000 kV/m in the atmosphere, electric discharge with light is observed following the generation of EM pulses. In nature, the discharge between lower clouds and the ground as well as among clouds is called lightning, which also generates EM pulses.

A discharge between clouds and the upper ionosphere surrounding the Earth is called a *sprite* and gives EM pulses with longer duration than lightning.

More charges are produced in the condenser when a dielectric material (mostly an insulating material) is placed between the electrodes, as shown in Figure 3.10 (b). The dielectric material is electrically polarized by the electric field, that is, the opposite charges are separated (as shown by the arrows), so more charges flow to the electrodes to cancel the polarization. These charges at the electrodes, called compensating charges are released when the polarization disappears. So, many charges are stored in the condenser. One could consider the ocean and ionosphere one large condenser, and the Earth's upper crust a dielectric condenser sandwiched between the conductive lower crust and the ocean.

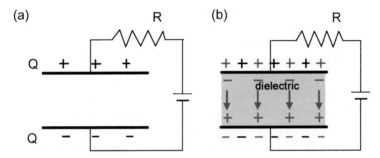

Figure 3.10 (a) Charges appear at the electrodes of a condenser by application of voltage. (b) Insertion of a dielectric material between the electrodes generates polarization (indicated by the arrows), and more charges (compensating charges, indicated by the grey + and - signs) appear at the electrodes to cancel out the polarization of the material.

3.3.6 Propagation of EM waves in the earth's crust and atmospheric waveguides

Two conductive planes reflect EM waves between themselves and constitute the simplest waveguide in which EM waves are trapped. There is a cutoff frequency below which EM waves decay drastically (exponentially) rather than travel along the waveguide. The cutoff frequency is the frequency at which the half-wavelength of EM waves is equal to the separation of the conductive plates.

The ionosphere (where the atmosphere is ionized by UV solar wind and cosmic radiation) and the conductive ocean surface, form a waveguide called the *earth's atmospheric waveguide*, with a separation of 60 km (between the ionosphere and the ocean) during daytime and 120 km at night. The cutoff frequency can be calculated, respectively, to be 2.5 kHz during the daytime and 1.25 kHz at night. Space weather such as solar flares can disturb the magnetosphere and ionosphere and cause magnetic storms resulting in disturbed transmission of EM waves. There are reports that earthquakes follow geomagnetic disturbances (Kormiltsev *et al.*, 2002).

The earth's crust between the ground surface and the conductive Moho-plane (or the Conrad plane which exists at the boundary of the upper and lower crusts) may constitute a dielectric slab waveguide called an *earth's crust waveguide* (Ikeya *et al.*, 1997b). This waveguide was found during the measurement of earth-origin EM waves using borehole antenna (Tsustui, 2002) as described in Chapter 11. The two earth waveguides are connected to each other through land and islands.

The electric field intensity of EM waves emanating from the slab decreases exponentially as a function of distance from the surface. Such EM waves are called *evanescent waves*. The two earth waveguides may be considered as condensers.

An experimental model of the earth's crust using a map with a scale of 1/1,000,000 was constructed (Figure 3.11) to study the propagation and interference of EM waves with each other, and hence to check whether EM waves could travel significant distances in the crust. Aluminium and salt water having a resistivity a million times less than in the natural environment were used to simulate the ocean and land, leading to a million times smaller skin depth using a million times smaller wavelength than that of SEMS. The maps of Japan and the west coast of California (used in the model) were commensurately reduced in scale (Sato *et al.*, 2001; Ikeya *et al.*, 2002). The model showed EM waves travelled significant distances in the earth's crust and were disturbed by fault planes.

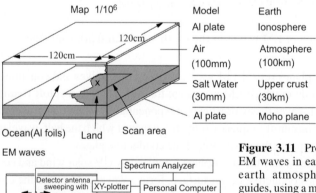

Figure 3.11 Propagation of EM waves in earth crust and earth atmospheric waveguides, using a model consisting of a million times higher conductivity for concordant skin depth.

3.3.7 Piezoelectric polarization and piezo-compensating charges

Piezoelectricity is the electric polarization produced in certain crystals and ceramics by the application of mechanical stress. Piezoelectric polarization occurs when the ions in a crystal are not symmetric at the center. There are many synthetic ceramics which show intense piezoelectric polarization when stressed. For example, the ceramic, *barium titanate*, has been used as a piezo-element.

A slight deformation produced by stress results in different center positions of negative and positive ions, as in Figure 3.12 (a) and (b). Thus, stress turns the mineral into an electric dipole with separated positive and negative charges. In Figure 3.12 (c) an oscilloscope shows the polarization (expressed in voltage variation) resulting from a hammer blow on a single crystal of quartz. Some automatic lighters in gas ovens and cigarette lighters employ piezoelectricity to generate electric discharges.

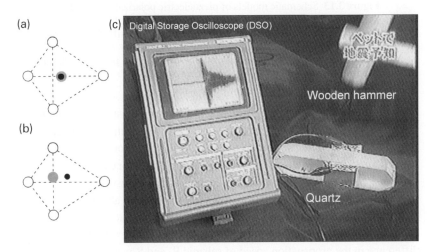

Figure 3.12 (a) An ionic crystal lattice in its natural state with asymmetric ion configuration. (b) the lattice is polarized by stress-induced distortion causing the centers of the positive and negative charges to deviate. (c) A digital storage oscilloscope shows electric charge generated by hitting a synthetic quartz crystal with a hammer.

Polarization charges are fixed within the mineral under normal conditions and can neither migrate in the material nor in the surrounding matrix minerals. However piezo-compensating charges can move and do so in an attempt to cancel the polarization charges. They do not physically contact the polarization charges, but remain close to them (See Figure 3.13). The outcome is a large number of charges in the rock, and the generation of significant voltages for the following reasons.

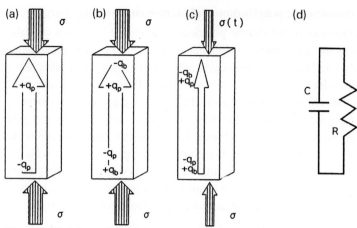

Figure 3.13 Schematic models of piezoelectric polarization and the production of piezo-compensating charges in granite in the earth's crust. (a) Stress-created piezoelectric polarization charge densities $+q_p$ and $-q_p$. (b) Piezo-compensating charges, $+q_b$ and $-q_b$ in a slightly conductive medium situate themselves near the polarized charges. (c) The piezo-compensating charges combine with the polarized charges on release of stress, and neutralize them. (d) an equivalent circuit.

Granite, which comprises the bedrock of the earth's crust, consists of piezoelectric quartz grains, insulating mica and relatively conductive feldspar. Normally there is little preferred direction to the quartz grains, and one would think that the various small voltages would cancel each other because for each one oriented (say) east-west, positive to negative, there is another oriented west-east, positive to negative. However piezoelectricity may still arise from small amounts of preferential (not-random) orientation of crystalline axes. This orientational anisotropy (as it is called) has been observed in the minerals pyroxene and olivine, because of the difference in the propagation speed of seismic P- and S-waves.

It may also arise where there is completely random orientation. To give an analogy, excited atoms in a fluorescent lamp are randomly oriented and each is an electric dipole. They do not cancel each other out. From the number of photons emitted from the excited atoms it may be calculated that the average polarization is proportional to the square root of the number. Similarly, completely randomly oriented quartz grains may emit EM waves from a polarization that is proportional to the square root of the number of grains.

Our experiments indicate that the production of charge is about 1% - 0.01% that of quartz for a "laboratory-sized" piece of granitic rock (Sasaoka *et al.*, 1998). Thus voltages may be produced. The degree of orientation is not known.

The idea of piezoelectricity as a cause of earthquake precursors has been abandoned by some geophysicists because they believe the rocks are too conductive and the charges will be able to reach and neutralize each other, giving zero voltage, but laboratory experiments show voltages may still occur in rock under stress (Figure 3.14). If there is a change in charge densities either in polarized or piezo-compensating charges through this stress, either by local microfractures of rocks causing dilatancy, or by faulting (main shock), EM waves may be produced.

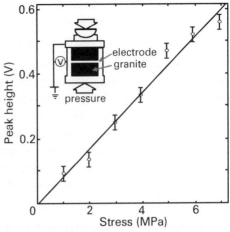

Figure 3.14 Voltages produced in granite as a function of stress. In insertion shows the apparatus (Sasaoka et al., 1998).

Following earthquakes we studied EM precursor phenomena based on piezo-compensating charges of completely randomly oriented quartz grains, but the calculated EM intensity was orders of magnitude too small to produce the phenomena. For that there would need to be some minimal alignment of quartz grains leading to alignment of piezo-compensating charges at stress release, which is very probable. However there are several other mechanisms for generating EM pulses during the fracture of solids which are not piezoelectric materials e.g. amorphous materials like glass also generate EM radiowaves, as well as light, upon fracturing. So other mechanisms might be in operation.

If a country has relatively little granite (e.g. New Zealand), there may be fewer opportunities for the development of these high charges, fewer EM waves and fewer earthquake precursors observable. This will be the case particularly when the focal depth is greater than about 10 km, resulting in attenuation (weakening) of the wave. A focal depth of 10 km is common for inland earthquakes in Japan.

3.3.8 Electrokinetics and other mechanisms of stress-electricity conversion

(a) Streaming potential (zeta potential of water)

Terada (1933) discussed electro-atmospheric effects due to the dipole field caused by the streaming potential in water flow through porous rocks and soils. Either positive or negative ions are adsorbed on the surface wall of minerals, while ions with opposite charges flow in the fluid. This forms dipolar charges depending on the fluid velocity and the concentration of ions (See Figure 3.15 (a) for a schematic representation). However, the electric field is too weak to cause electric breakdown of the atmosphere sufficient to explain earthquake light (Mizutani *et al*, 1976).

(b) Crack electrification: Lenard splashing

Large charges are produced by mechanical dispersion of pore water into droplets. The mechanism of charging droplets by splashing or spraying was termed *waterfall electrification* by Lenard. Vapor with a relatively large surface is charged negatively, while a large droplet is positively charged. It is possible that vaporization of pore water by sudden destruction of the surface layer during local faulting could generate intense charges responsible for atmospheric discharges (Lockner *et al*, 1983). This vaporization might occur in the opening tip of microcracks resulting in an ensemble of electric dipoles in the microfracture zone [See Figure 3.15 (b)].

(c) Charged dislocation by crack formation and exoelectron emission

Ionic crystals have an innate line defect called dislocation, which is charged locally and neutralized by charged impurity ions, also known as an ion impurity cloud [Figure 3.15 (c)]. Quick movement of the dislocation by stress deformation results in polarization until the impurity ions move slowly to compensate the charge. Crack formation and propagation involves the motion of these charged dislocations and has been suggested as a physical model for the formation of electric earthquake precursors (Slifkins, 1993; Tzanis and Vallianatos, 2001, 2002).

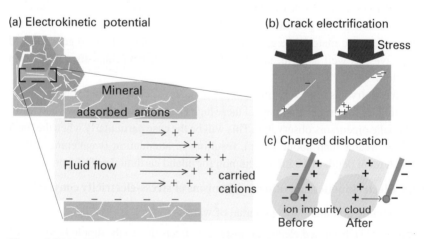

Figure 3.15 (a) Electrokinetic potential formation from fluid flow into a crack in fractured mineral rock. (b) Crack formation and propagation as a mechanism of conversion of stress to electricity and (c) Left: A negatively charged dislocation line with positively charged impurity clouds. Right: Movement of the dislocation, forming an electric dipole and leaving charged impurities.

Electron emissions from traps filled by natural radiation (called *exoelectron emissions*) occur through crack formation (Enomoto *et al.*, 1996). Rabinovitch *et al., (1999)* studied EM radiation during uniaxial and triaxial compressions of granite, rhyolite, chalk, glass and ceramics, and produced a Gutenberg-Richter type relation of EM intensities to fracture size.

So, apparently, a mechanism, or mechanisms other than piezoelectricity are possible for the generation of EM pulses, though the actual means of EM generation by fractures in solids is still unclear.

3.3.9 Frictional electricity: Van de Graaff electrostatic generator

One can be surprised by an unexpected spark and shock from static electricity upon touching a metal doorknob or unlocking the door of a car. This is because static electricity has built up in the body e.g. a charge of one millionth of a coulomb accumulates when a person walks on a carpet, and is released (giving a shock) when he or she touches an electrically grounded object like a doorknob.

If two materials are rubbed against each other, one is charged positively and another negatively. This is due to a static electric charge produced by friction of two materials. The order in which the materials rubbed will become positively charged by transferring electrons is: glass> hair> mica> nylon> wool> rayon> silk> paper> amber> copper> rubber> Teflon. For example, if glass is rubbed against rubber, the glass will be positively charged and the rubber negatively charged. Some electric discharge phenomena such as lightning at the time of landslides are thought to result from discharges induced by friction of minerals.

A *Van de Graaff generator* [schematically shown in Figure 3.16 (a)] is an electrostatic generator which uses frictional electricity to produce a high voltage. It consists of a large hollow metal dome mounted on a hollow insulating support and base. A rubber belt runs through the support from a lower pulley in the base to an upper pulley within the dome. A belt, stretched between two pulleys, is either motor or manually driven. The lower pulley is covered with felt and the upper one coated with a thin film of cellulose nitrate (collodion). Driving the belt from the lower pulley strips electrons away from the felt and attaches them to the belt. The belt carries the electrons upward to the upper pulley and is charged positively there by friction with the collodion film, leaving more electrons than it carried from the lower pulley. Electrons move from the upper pulley to the metal dome through a metal comb close to the upper pulley. Negative charge can be accumulated to a voltage as high as 250,000 V for a dome diameter of 25 cm (which corresponds to an electric field of one million V/m at the dome's surface).

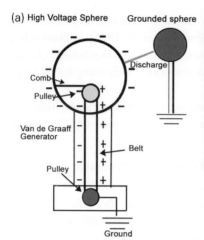

(a) **High Voltage Sphere**   **Grounded sphere**

Figure 3.16 (a) A schematic drawing of a Van de Graaff electrostatic generator using frictional electricity and (b) arc discharge from the generator to a grounded metal sphere on the right.

A grounded metal sphere, shown to the right in Figure 3.16 (b), is used to attract negative charge from the dome by an induced positive charge on the grounded sphere. An intense electric field is produced between the dome and the sphere. By bringing the grounded sphere close to the dome an intense electric field of about 3 million V/m can be produced, causing an atmospheric breakdown which creates a bright bluish-white arc accompanied by pulsed EM waves. This is called *arc discharge*.

At a much less intense electric field of 100,000 V/m, a continuous EM discharge with a faint glow occurs. This is called *corona discharge* and is used in laser printers and copy machines. Corona discharges generate ions and an ion wind. An insulator or a material isolated from the ground is charged up when it is exposed to ion winds.

A reversal of polarities occurs if negatively charged Teflon is used for the lower pulley or the lower pulley is coated with collodion.

3.3.10 Absorption ratio for a dielectric sphere

The scattering of optical light as small particles is known as Mie scattering [Figure 3.17 (a] and can be calculated using Maxwell's equations. Conversely, the absorption of EM waves by a sphere can also be calculated using Maxwell's equations. The absorption rate of EM waves depends on the size and optical properties (the dielectric constant and conductivity) of the spheres and the frequency of the EM waves. When the incident EM waves (light) are fully absorbed by a sphere, the absorption ratio is 1.0. The absorption ratio is strongly dependent on the EM wavelength (or frequency).

The absorption rate of EM waves at several frequencies by a saltwater sphere simulating animals is shown as a function of radius up to 100 cm in Figure 3.17 (b). The rate peaks at the frequency at which the wavelength (obtained by dividing the

Figure 3.17 (a) Mie scattering for a sphere of salt water. (b) The absorption ratio of EM waves as a function of the radius of the salt-water sphere at several frequencies. The cross-section ratio at 50 and 60 Hz at the size of about 1 m is almost zero (10^{-27}).

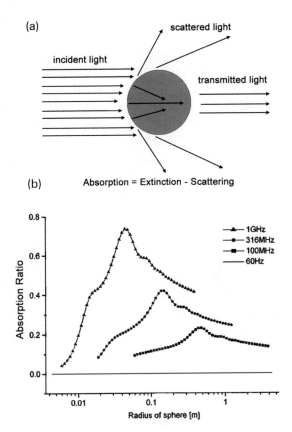

speed of light by the frequency) is almost equal to the circumference of the sphere. For example, a frequency of 1 GHz corresponds to a wavelength of 30 cm. A sphere with a radius of 5 cm having a circumference of 31.4 cm shows a peak absorption rate at 1 GHz (wavelength 30 cm). Therefore, animals of that size might absorb EM radiation at 1 GHz depending on their body resistivity.

3.3.11 Electric field effects on animals

Many aquatic animals such as catfish and sharks have electrosensory organs and are extremely sensitive to minute electric fields of 0.001 mV/m. They use this electric field sensitivity to capture prey hidden beneath mud and sand. Sharks have attacked fish in a chamber blocking all but weak electrical stimuli of a millionth of a volt. An artificially produced electric field elicited the same attacking behavior. Biological electromagnetism is described in Chapter 4, Unusual Animal Behavior.

Land animals might also perceive electric fields of both static and EM waves through wet tongues, eyes, and paws with high conductivity acting as antennae, and body hairs acting as electrostatic mechanoreceptors. Only antennae that resonate at the frequency range being emitted can efficiently detect those EM waves. Hence emitter and receiver antennas should be similar in size and shape. A window of communication at a particular frequency range may therefore be opened to animals of the same species, the frequency range depending on the size of the animal.

Electric fields affect synapses, which connect nerves for information transfer within the nervous systems of animals. Birds flying or fish swimming in the earth's magnetic field, detect electric fields whose intensity is proportional to their velocity. In their natural environment fish are easily able to orient themselves east or west by detecting what is happening to the electric field inside them. If animals can detect an electric field of ten millionths V/m, as some fish with electrosensory organs can (e.g. sharks and catfish), they are detecting an electric field rather than the Earth's magnetic field. That is, the animal responds to the internally induced current produced by the electric field rather than to the magnetic field itself.

The absorption ratio of EM waves by an object is somewhat analogous to the impedance of alternate current (AC), corresponding to direct current (DC) resistivity described in Section 3.3.1 (d). Resonances occur at particular frequencies depending on the size and shape of the object. If the frequency is the microwave frequency of 1 GHz, for example, an animal 10 cm long absorbs it quite intensely, but an animal twice as big is less affected. If animals respond to EM waves associated with preseismic events, then some may be greatly affected by EM waves while others of different sizes and small conductivities may not sense them at all. The impedance (electrical resistance) of an object also differs depending on the direction of the electric field. Hence there is directional dependence in addition to size (frequency dependence) in the effects of EM exposure on animals. This seems to be rarely understood by biologists, leading to irreproducibility in their experiments.

3.4 Summary

This chapter has introduced the reader to enough physics to understand the following chapters. The EM model of a fault draws particularly on the behavior of piezo-electricity and piezo-compensating charges, although other processes: dilatancy-fluid diffusion and mechanisms of stress-electricity conversion, are also considered to be at work.

Big earthquakes occur when giant tectonic plates grind against each other and also when this pressure forces geological faults and fault systems to release their accumulated stress, producing seismic waves. But well before the main fracture, increasing stress on rock creates microfractures. Our thesis is that this process generates electromagnetic (EM) signals that cannot be detected by modern seismographs but which possibly explain reported earthquake precursor phenomena—although the detailed mechanism by which this occurs is not yet clear.

The following chapters attempt to test this premise scientifically and to evaluate the intensity of the signals—quantitatively rather than qualitatively—from reported unusual phenomena before earthquakes.

Unusual Animal Behavior: I
What do they Detect?

Electric Field Effects

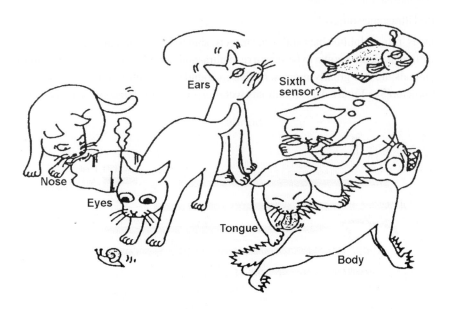

There are five animal sensory organs: the eye (light), the nose (smell), the ear (sound), the tongue (taste) and the skin (touch, wind, temperature). Buddhism teaches that these five senses and a so-called sixth sense are color (i.e. reality) and that color is empty and emptiness is colorful (A. Hochi).

4.1 Introduction

In legends and modern reports of animal behavior before earthquakes, animals are said to behave in unusual ways. Readers might glance back at Chapters 1 and 2 and at Table 2.1, to reacquaint themselves with the range of reported behaviors—usually of a kind showing uneasiness, anxiety, distress or panic.

If animals really are sensing something before earthquakes that makes them uneasy or fearful, what might they be sensing? How might they be sensing it?

This chapter will look at signals animals could be detecting that might produce anxiety or fear, and describe the results of laboratory or field experiments to test animal responses to stimulation by EM waves, like those produced by lightning, changes in the weather and before earthquakes.

4.2 What do animals detect?—what are their sensors?

4.2.1 Sounds (acoustic waves) and vibration (foreshocks)

Are animals sensitive to:

(a) Ultrasonic sounds?

Dogs, rats, dolphins, bats and some insects, said to behave in unusual ways before earthquakes, are known to be sensitive to ultrasonic sounds. However, most of these animals are not greatly more sensitive to sounds at a high frequency range than humans (See Figure 4.1). If animals were responding to ultrasonic sounds before

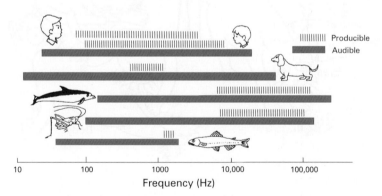

Figure 4.1 Frequencies that animals can hear and in which they can produce sound. Humans have a wide range (except for the extreme high and low-frequencies). Insects and fish are not sensitive to low-frequency sounds (From the *Asahi Encyclopedia for Elementary School Pupils*).

earthquakes, humans would also be reporting them more often. Furthermore, ultrasonic sounds, which might be produced by microfractures in underground local rock, cannot reach the surface because of rapid attenuation during travel. Animals would only detect sounds from a depth of 10 m at most (Buskirk, 1981).

(b) Subsonic (infrasonic) sounds?

Low-frequency sounds generated by large-scale rock fractures travel a long distance. Modern seismographs are only capable of detecting 10 - 100 Hz P- and S-waves, but their records show they travel to the other end of the Earth. It is said that whales can communicate long distances with low-frequency sound. Could heightened sensitivity to subsonic sound explain odd animal behavior before large quakes?

It seems not. Human beings can hear a considerable frequency range—from 10 Hz to 20 kHz—and so should be able to hear at least some of the same low frequency sounds animals are hearing. But there are no reports of humans hearing low frequency sounds a week before quakes at the time some animals are deserting these areas, though many people say they hear low frequency sounds just before the arrival of P-waves.

Dolphins, insects and some fish, for example, are less sensitive to subsonic sounds than humans (See Figure 4.1), but they are reported either to have behaved in unusual ways or to have fled from the epicenters of earthquakes days before the arrival of P-waves.

(c) Seismic P-waves preceding tremors?

Earlier generations of Japanese used pheasants to forecast earthquakes because they gave short warning cries beforehand. Omori (1923) reported that pheasants respond to P-waves and twitter before the arrival of S-waves. The early warning systems of UREDAS (Japan Rail) and CUBE of Caltech & USGS (Section 3.2.2) also detect P- and S- waves, and a warning system that shuts down electricity and gas supply in private homes on detection of P-waves is under study.

Fish and animals are potentially able to detect earthquake precursor sounds. The lateral-line system in fish is composed of receptors in pores and canals on fins, at a line along the length of the body. It responds to accelerated water motion over the surface of the skin and detects sound waves created by that motion. The well-known otolith organ of the inner ear responds to the acceleration of animals as a whole. Both could potentially detect precursor sounds.

However, most reported unusual animal behavior occurs well before the arrival of P-waves, that is, a few hours or a few days before any detectable foreshocks in seismographs. So it is difficult to ascribe unusual animal behavior to perception of sounds accompanying P-waves.

(d) Sounds and vibrations?

Octopuses were reported moving about on land in a clumsy topsy-turvy way before the Kobe Earthquake. Could octopuses have a heightened sensitivity to sound waves or vibration? Five octopuses subjected to loud orchestral music at a live concert attended by the author were unaffected by it. While the water in the aquarium trembled with the vibration of the drums and instruments the five octopuses remained still. Intense resonating sounds and vibrations did not affect them.

Some fish are certainly very sensitive to sounds, and leap out of the water in response. But they adjust to them very quickly and the leaping behavior soon stops. Again, because fish and humans can hear roughly the same frequencies there would be more reports of people hearing sounds well before earthquakes if animals were responding to sounds and vibrations before earthquakes.

4.2.2 Odor or outgassing before earthquakes

Detection of smell enables animals to find food, prey, enemies and mates. Some animals can even detect a single molecule of odor-causing chemicals. Dogs, which have a hugely keener sense of smell than humans, are used by airport customs officials to find hidden drugs in baggage. So, could odor or outgassing from the ground cause unusual animal behavior?

Microcrack formation before earthquakes may release gases such as the radio-active gas radon, methane, nitrogen oxides, sulfur oxides and hydrogen sulfide, some of which have been studied for use in short-term earthquake prediction. In high enough concentrations these could make animals fearful and panicky, but the concentrations are not high enough. Some chemical reactions before earthquakes might generate toxic gases causing the same reactions. Nitrogen oxides, for example, might be generated by electric discharges in the earth, as by lightning in thunderstorms, but, again, not in high enough quantities to cause panic among animals. Although the hypothesis of earth gas and odor as a cause of unusual animal behavior remains, it is difficult to prove or disprove.

4.2.3 Infrared or far infrared light

Could unusual sensitivity to infrared light be a factor in unusual animal behavior? Snakes, for example, have infrared detectors and use them to capture prey. Some people believe that infrared light emanating from an earthquake epicenter deep in the Earth's crust, can raise the local temperature, warming snakes so that they wake up and come out in winter, believing it to be spring. Although snakes can emerge early in spring from hibernation, large numbers of them have been observed doing so in midwinter before large earthquakes. Some say far-infrared light has beneficial effects, penetrating the body, warming it and enhancing metabolism. But this should not alarm animals. Something else appears to be turning a comfortable winter re-treat into an uncomfortable environment. Cats in Japan like to sleep under infrared rays from heaters in winter, but their reported behavior before earthquakes is to meow plaintively and run out of the house carrying their kittens with them.

Although some animals, such as butterflies, detect light at unusual wavelengths, the range of light wavelengths seen by animals is well within those seen by humans. So it is difficult to believe that so many animals, behaving in peculiar ways before earthquakes are doing so because they are detecting light undetectable by humans.

4.2.4 Magnetic fields

(a) Do animals detect magnetic fields?

Migration of birds for hundreds of kilometers has sometimes been ascribed to their sensitivity to the Earth's magnetic field, as has the long-distance navigation of some vertebrates. Magnetic field effects are suspected in the case of bacteria, honey bees and homing pigeons. In their tissues there are small natural needles of magnetic mineral, magnetite (Fe_3O_4), which the organisms have synthesized themselves. The energy for the magnetic interaction of this magnetite with the earth's magnetic field is ten million times stronger than for other biological substances. There is evidence that fin whales respond to the geomagnetic field during migration (Walker *et al.*, 1992, 1997) and rainbow trout sense magnetic fields of similar strength to the earth's magnetic field. The unusual behavior of homing pigeons before earthquakes has been ascribed to a change in the earth's magnetic field.

Magnetic fields as a cause of unusual animal behavior were discussed in the United States

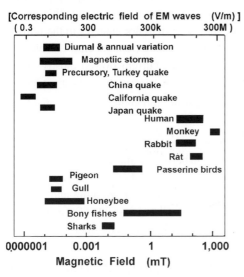

Figure 4.2 The intensities of magnetic field that animals can detect (Ruth *et al.*, 1980). The corresponding electric field for EM waves has been added in the upper scale for later discussion. Differences between voltages in this Figure (and in Figure 4.4) and the experimental results in this chapter, are probably due to our use of pulsed electric fields and the exclusive use of static fields by Ruth *et al.*

Geological Survey (USGS) Symposium in 1979 and the sensitivity of various animals to magnetic fields in the unit of mT (a thousandth of one Tesla) was reported (Figure 4.2). Data are for the range of a static magnetic field. In the upper scale, the corresponding electric field of EM waves has been added for later discussion. Kirschvink (2000) also ascribed unusual animal behavior to magnetic field effects before earthquakes.

Magnetic field sensing by animals is still controversial. Some studies indicate that pigeons do not make use of magnetic cues on their flights. Pigeons which have been fitted with modern, intense, permanent magnets producing magnetic fields a thousand times more intense than the earth's magnetic field, still fly home without difficulty (Sheldrake, 1994). They appear instead to judge their direction from the position of the sun using an internal clock which regulates circadian rhythms.

Five octopuses in an aquarium showed no change in their behavior during application of an external magnetic field. During an open lecture the author held an intense, permanent magnet (Neomax; Nd-B-Fe alloy) generating a few tens of mT near the octopuses from outside the aquarium and then moved it quickly so that they felt an abrupt change in the magnetic field. But they remained still. This simple experiment showed that neither a static magnetic field nor changes in the Earth's magnetic field affects octopuses. Changing a magnetic field generates an electric field and eddy current. However, in this case, the current flowed mostly in the conductive salt water in the aquarium and very little in even the nearest octopus.

(b) Magnetic field changes before earthquakes are too small to cause behavioral changes

Similar experiments were performed on many animals, but produced no behavioral change. Changes reported in the magnetic field before earthquakes are less than one ten thousandth of the earth's magnetic field of 0.03 - 0.04 mT. Some animals may use ferromagnetic bio-minerals in their brain or body for orienting and homing using the direction of Earth's magnetic field, but will be incapable of responding to a variation less than one ten thousandth of the Earth's magnetic field.

(c) Do human beings feel magnetic fields?

One report claimed that some American miners could detect nearly one millionth of the earth's magnetic field. J. Taylor (1980), a physicist, in his book *Science and the Supernatural*, speculated they might sense small changes of magnetic field. But his own experiments indicated that miners did not even detect a magnetic field of 50 mT—one thousand times the earth's magnetic field.

People are sometimes exposed to an extraordinarily large magnetic field of the order of a few Tesla when they go to hospital for MRI scans (body scans by computerized magnetic resonance imaging). They usually feel no magnetic field unless they move their heads quickly under an intense magnetic field; body movements under a magnetic field induce an electric field and an electric current. But the magnetic field variation before earthquakes is much less than several millionths of field produced by MRI scans. So changes in the Earth's magnetic field before a large earthquake are very unlikely to produce unusual animal behaviors.

4.2.5 Electric fields

(a) Electric fields

Earth and space have different electric charges and this creates voltages in the atmosphere. At sea level there should be a voltage of about 100 - 200 V between a person's nose and toes . However, this voltage is not actually present in the animal since animal tissues conduct charge. An animal's head has the same charge as the Earth and the foot has the opposite charge, but the electric field in the animal cancels the external electric field, so current does not flow [Figure 4.3 (a)]. On the other hand one would be killed if one's head touched a high-amp power line while the toe touched the ground, because a large current would then flow through the body from the power line. It is the internal electric field and flowing current in a body that affect animals [Figure 4.3 (b)]. This is hardly understood by the general public and even some scientists are confused about it.

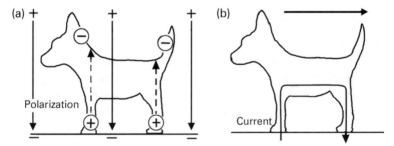

Figure 4.3 (a) A vertical electric field without current flow does not affect animals because internal polarization cancels the field. (b) Horizontal alternating and pulsed fields induce electric currents in the body depending on body impedance/resistance and the frequency of the alternating current.

Pioneering studies on the effects of electric stimuli on catfish after the Great Kanto (Tokyo-Yokohama) Earthquake in 1923 (M7.9: 142,000 casualties) suggested that telluric electric signals (actually electric fields created by regions of differing electric potentials in the earth) generated before earthquakes might elicit unusual behavior from aquatic animals (Hatai and Abe, 1932; Kokubo, 1934; Abe, 1935). Fish might sense variations in electric fields or telluric currents (flowing in water)

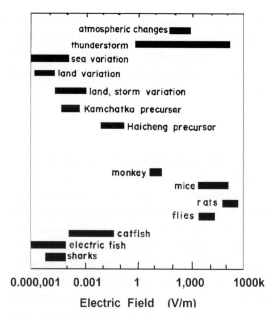

Figure 4.4 The range of electric field intensity that animals sense (Ruth *et al.*, 1980)

before earthquakes. Electric fields as a cause of unusual animal behavior were also discussed at the U.S. Geological Survey (USGS) Symposium Proceedings at Menlo Park in 1979. Although the data are old, the relevant figures are reproduced in Figure 4.4.

Work in the early 1990s showed that some aquatic animals have an extraordinarily sensitive electrosensory system for capturing prey and communicating with each other (Bastian, 1994). A voltage as low as 0.04 mV/m is perceived by sharks through their extremely sensitive electrosensory organ *Lorenzini's Ampulla* (discussed in Section 5.2.2 and Figure 5.2).

(b) Electric fields induced by movement in magnetic fields

Electric fields are induced by tidal flows in the Earth's magnetic field. Some fish can detect the electric current produced by their own movement in a magnetic field and orient themselves by it (passive electro-orientation). If the speed of movement is 1 m/s in a magnetic field of 0.04 mT, an electric field of 0.04 mV/m is produced following a formula:

(the velocity in meter/s) x (the magnetic field intensity in Tesla) = (the electric field in volts/meter).

Some aquatic animals can detect an electric field of this intensity (about 0.05 mV/m) with their electrosensory organs, and so, if moving, can sense the direction of the Earth's magnetic field through the action of the electric field on their nerves (active electro-orientation). This sensitive orientation system could quite conceivably be disturbed by electric pulses generated by underground microfractures of rocks from accumulated seismic stress.

4.2.6 Charged aerosols (The Serotonin Syndrome)

Aerosols consist of small dust-like particles of about 0.1 - 0.0001 mm, so small that they remain suspended in air just like molecules moving around in solution, in contrast to those in a solid, which stay in place. Aerosols are familiar as the air polluting suspended particulate materials (SPM) in combustion gases. If aerosols have positive or negative charges, they are called *charged aerosols*. Small negative ions are said to have biologically healthy effects, and large positive ions hazardous effects.

Tributsch (1978, 1982) suggested that unusual behavior in land animals was caused by charged aerosols emanating from the epicenter and that their neutralization caused earthquake light. He suggested land animals had a metabolic reaction to charged aerosols and manufactured increased amounts of serotonin (a neurotransmitter affecting mood). He called this response, the *Serotonin syndrome*.

The ionization of air by radioactive radon released from the ground in advance of certain earthquakes also produces an aerosol and increases atmospheric conductivity. Land animals may be sensitive to charged aerosols in the air because a current would then flow from the charged-up body to the ground, maybe producing unusual electrophysiological responses.

However, aerosols quickly diffuse away in wind. Also, fish and reptiles in small aquaria and animals in modern insulated buildings with much reduced aerosol levels have behaved in unusual ways before earthquakes, making aerosols an unlikely cause.

4.2.7 EM waves (< 10 kHz): electric pulses (> 0.1 ms)

Electric fields produced by EM pulses generate current in animals. There is a specific frequency range of EM fields that exerts the maximum effect on animal tissues. An electric pulse is composed of continuous EM sine waves at different frequencies (See Figure 3.9) which goes by the interesting name, *wave-packet of EM waves*. Some EM waves composing the wave-packet may induce electric currents in an animal body at the right frequency range where body impedance is small.

Tributsch (1978) abandoned the hypothesis that EM waves affect animals because he heard that cattle and horses in what he believed to be electrically-shielded livestock wagons still showed unusual behavior before an earthquake. The wagons were built of sheet iron with several rectangular openings. However, EM waves may still have penetrated through the rectangular openings in the wagons, simply because one is able to listen to radio broadcasts and use cellular phones in such wagons. In fact the wagon might even become a resonator of radio waves, according to the known physical principles of resonance. So it is still possible EM waves cause unusual animal behavior.

Taylor (1980) tested himself by applying an intense electric field to his body (100 - 1000 times greater than the earth's electric field) and could not find any effect at 1 MHz. A very fast AC electric field, such as Taylor used, changing its polarity a million times a second, produces an extremely small electric current (because of impedance), but it is probably not detectable and its direct action on nerves is unlikely because the basic reaction time at a nerve synapse is slower than 0.1 millisecond (ms). To feel any effect Taylor would have had to have used a frequency lower than 10 kHz or modulated the power at a lower frequency than 10 kHz. We shall demonstrate the effect of pulsed electric fields in the following sections.

4.3 Electric field experiments on aquatic animals

4.3.1 Experimental conditions

(a) Electric field experiments

Metal electrodes such as stainless steel nets or aluminum (or copper) plates were immersed in an aquarium where aquatic animals were kept. Electric pulses were applied using a series of 1.5 V batteries by switching a DC electric field on and off manually. The pulsewidth, monitored using a digital storage oscilloscope, was about

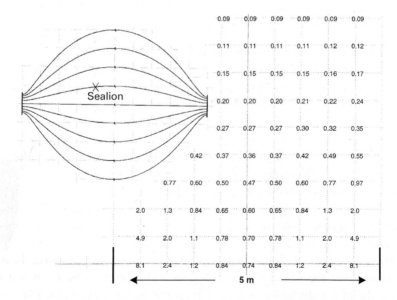

Figure 4.5 Calculated voltage at each mesh site for an electrode separation of 5 m and a voltage of 100 V (an apparent intensity of 20 V/m) in a pool, using software which simulates the electric field intensity (Ikeya *et al.*, 1997c).

0.05 seconds (50 ms). A pulse generator was also used with a pulsewidth of generally 5 ms and the repetition rate was either 10 - 100 pulses/s or 1 pulse/s. An electronic massage apparatus, Elepulse (Omron HV-F04), was used in the field study.

The electric field intensity was obtained basically by dividing the applied voltage by the electrode separation for large electrodes with a small separation, or by using an electric field simulator (a computer program) to calculate the field when separation distances were large. Figure 4.5 shows the numerical values of the calculated electric field intensity when the applied voltage is 100 V with an electrode separation of 5 m. The behavior of animals upon application of voltages was photographed using a commercial video recorder.

(b) EM pulses generated by electric discharges

Arc discharges using a Van de Graaff generator can generate EM pulses of 60 V/m with a time range of a microsecond. These were measured to have a peak voltage of less than 1 V/m in the water of an aquarium, using parallel plate electrodes and a digital storage oscilloscope.

(c) Exposure of an aquarium to charged aerosol

Wadatsumi (1998), who follows Tributsch's hypothesis, exposed the water surface of an aquarium in which a catfish was kept to charged aerosol. He found the catfish moved suddenly. But this would be in response to induced current flow in the water, not by direct exposure to the charged aerosol. Since the aquarium, which is isolated electrically, is charged up by exposure to charged aerosol, a pulsed current may flow in the water by natural discharge through air to the ground.

4.3.2 Catfish and eels: Which is more sensitive?

Electric field effects on two catfish—whose normal behavior is resting motionless, as in Figure 4.6 (a)—were studied by applying electric pulses. It is known that

Figure 4.6 Photographs of catfish: (a) motionless in a pipe and (b) in violent motion at an applied electric field of 4 V/m.

catfish with sensitive electrosensory organs can perceive an electric field of 0.005 mV/m, a minimum intensity gained by gradually reducing the applied voltage. Their fins and whiskers moved with each pulse at 2 V/m for a pulsewidth larger than 0.1 ms; their muscles twitched and they moved violently at 3.5 V/m. Repetition made them panicky as shown in Figure 4.6 (b). Eels, in a polyvinyl chloride pipe placed in a plastic aquarium, moved suddenly at a nearby arc discharge from a Van de Graaff generator of about 1 V/m in water (as recorded by a digital storage oscilloscope). They rapidly came out of the pipe and quickly hid under the sand at an applied electric field of 0.5 V/m, which sent them into spasm. The catfish, although they might have felt the pulse, stayed still during equivalent discharges.

The catfish story told in the *Ansei Chronicle* in Section 1.3.3. may be given the interpretation that, since eels are agitated by lower electric fields than catfish, they might have escaped or hidden well before the earthquake, leaving the catfish behind. Although catfish can sense an electric field down to 0.005 mV/m (verified in various experiments) they are not perturbed by them. This is consistent with catfish behavior with prey; although their electrosensory organs detect their prey's low electric fields, they remain still but primed for action. Violent movement only begins at about 3 or 4 V/m. A field of similar intensity may have appeared in water before the Ansei Earthquake.

4.3.3 Freshwater aquatic animals: shrimps, crabs and alignment in Japanese minnows, guppies, and loaches.

Japanese minnows moved quickly when they were subjected to an electric field of 10 V/m (0.2 A/m^2) and aligned themselves perpendicular to the field direction as shown in Figure 4.7. Some were paralyzed at 50 V/m (1 A/m^2), but soon recovered. They responded to a pulse with a width of 0.1 ms, but were unaffected by shorter ones.

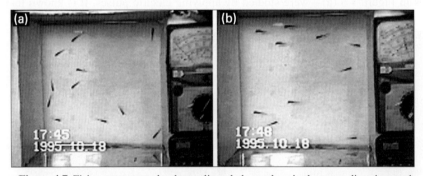

Figure 4.7 Fish were reported to have aligned themselves in the same direction, and to have died before the Kobe Earthquake. Japanese minnows (a) before and (b) after application of a pulsed electric field. The minnows aligned themselves perpendicular to the applied electric field. (Video photo.)

Guppies and loaches responded with quick and slow movements respectively, at the same current density. An applied pulse field of 5 - 10 V/m (0.1 - 0.2 A/m^2) with a width of 1 ms caused them to align themselves perpendicular to the electric field direction.

Freshwater shrimp (Astacidea procambarus), which were kept at the Takatsuki Aquapia at the fault end of the Arima-Takatsuki tectonic line, leapt out of the artificial channel in which they were being farmed before the Kobe Earthquake. The channel water was drawn from a well located close to a river. In the experiments carried out at the channel, electric pulses of 5 V/m attracted the shrimps to the electrodes and they fought each other in the excitement caused by the electric pulse stimuli.

Crabs jumped around and swam in an aquarium at an electric field of 10 V/m. Seafloor-dwellers, they have also been observed swimming on the surface and have been caught in fishnets before earthquakes, presumably because of their attempts to avoid intense electric fields on the sea floor.

4.3.4 Saltwater fish: lobster, squid, octopus, flounder and sea bream

Electric pulses with a width of 1 ms were applied to salt water fish in an aquarium and varied responses were observed; a pulsewidth less than 0.1 ms produced no effect.

Lobsters moved their legs slightly in response to every pulse at 5 V/m. At 10 V/m their bodies cramped and with each pulse above 15 V/m they jumped to the surface of the water in the aquarium. Consistent with this, Rikitake (1976), reported that lobster and cuttlefish normally living on the seabed were captured in fishing nets on the surface off the coast of the Japan Sea before the 1927 earthquake (M7.5).

Squid: This nervous and sensitive animal did not show a clear response up to 10 V/m. At 18 V/m its body color changed slightly to red and the eye became brighter by bioluminescence (See color plates). A large number of squid were caught before the Kobe Earthquake.

Flounder: A flounder did not show any response at 5 V/m. It began to move its fins at 10 V/m, above 16 V/m it moved its fins vigorously and its whole body violently. A flounder was observed leaping out of an aquarium a day before the Kobe Earthquake.

Sea bream: A sea bream spasmed at every pulse of 5 V/m, and at each pulse above 10 V/m, moved violently, then appeared stunned. Sea bream, captured in more than usual numbers before the Kobe Earthquake (Figure 2.14), were the most sensitive of the examined saltwater fish.

Octopuses did not move at a pulsed electric field of 2.5 V/m, but two of five octopuses began to move at 5 V/m. All of them spasmed with each electric pulse of

7.5 V/m and began to swim rapidly in panic at 10 V/m. They appeared less sensitive than catfish but more sensitive than flounder and squid.

4.3.5 Reptiles: Turtles and frogs

Turtles: A hibernating turtle woke up and attempted to climb an aquarium wall before the Kobe Earthquake (Wadatsumi, 1995). Turtles blinked at an applied electric field of 2 V/m. They stood up on hind legs to remove themselves from the shallow conductive water in the aquarium and climbed over other turtles to avoid the electric field stimuli as shown in Figure 4.8. They ran rapidly in panic at an increased voltage of 10 V/m.

Figure 4.8 Turtles attempting to avoid the electric field. To do so they (a) stood up trying to get out of the water, (b) tried to maneuver around the field (c) climbed over other turtles. (Video photo.)

Frogs: The respiratory rate of frogs in a small aquarium greatly increased when an electric field of 1 V/m was applied using electrodes immersed in water. They began washing their faces at 10 V/m—as cats did—and moved to the edge of the tank where the field was less intense. At 12 V/m they jumped even more rapidly to the edges of the tank.

4.4 Electric field experiments on land animals

4.4.1 Experimental conditions

(a) Direct Current (DC) electric field effects

Animals were placed between parallel plate electrodes to which a DC voltage of 3 kV was applied so as to produce an electric field of 10 kV/m. No unusual behavior was observed. This is because the current flows only briefly in the body, the internal electric field quickly canceling the polarizing effect of the external electric field (Figure 4.3).

(b) Electric pulse experiments (50 ms pulsewidth by quick on-off voltage)

In an experiment on ground electric field effects on animals, the animals were placed on a wet cloth that covered either the floor of their cages (land animals) or the perch rod (sparrows and parakeets). Electric voltage was applied by placing metal plate electrodes or aluminum foil electrodes at both ends of the wet floor or the perch [See Figure 4.9 (a)]. Switching the DC voltage on and off quickly generated pulses with a width of about 50 ms.

The surface electric field was obtained by dividing the applied voltage by the separation distance of the electrodes. The leg-to-leg voltage was estimated from the leg-to-leg distance and the electric field, and the critical body current was estimated by measuring the leg-to-leg (paw-to-paw) resistance with a tester (Ikeya *et al.*, 1996b).

(c) EM pulses by discharges using Van de Graaff and Wimshurst generators

(c-1) Van de Graaff generator

The discharges using a Van de Graaff generator produced pulsed EM waves from the high voltage sphere more intense than 10 W/m^2 (63 V/m) at an animal distance of 0.5 m, and 8 V/m at a distance of 3 m, as measured with a power meter.

(c-2) Wimshurst generator

Discharges to generate EM pulses were made using a Wimshurst generator, which generates electrostatic charges to two Leiden pots by manual rotation of a plastic

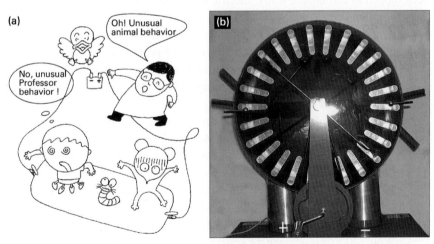

Leiden Pots storing static charges

Figure 4.9 (a) One student's perspective on electric field effects on land animals, using a wet carpet (S. Takaki). (b) A photograph of a Wimshurst generator that generates charges by friction and stores them in Leiden pots.

wheel [Figure 4.9 (b)]. The electric field intensity was varied by placing the animals at different distances from the generator. This apparatus is convenient for field experiments on the effects of EM pulses on animals where no electric power is available. However its frequent electric discharges sound like whipcracks, making it difficult to determine whether the animals are responding to EM pulses or the sounds of the discharges.

(c) Humane animal experiments

All animal experiments were carried out in collaboration with biologists operating within the parameters of animal protection regulations. Electric field experiments did not harm the animals because few batteries were used and voltages were only gradually increased upwards until effects were felt. The applied voltage was not increased beyond that causing slight muscle spasms of the kind produced by an electric massager. No animals were killed or injured in these experiments. Upon completion they were kept as domestic pet animals or returned to biologists for use in other experiments.

4.4.2 Mammals

(a) Human sensitivity to electric fields

Electric baths used in health care: What is the minimum intensity of an electric field that humans can detect? At 10 V/m the hairs of the feet can feel as if they have been dipped in carbonated water. Electric baths used for muscle massage are legally regulated at 20 V/m.

Barefoot sensing of telluric current: Usually, we wear shoes and do not feel telluric current. The Internet carried a report of a self-proclaimed Indian earthquake predictor who claimed to have felt an electric shock when he stood barefoot on the earth before the 2001 Gujarat Earthquake. The author did not feel up to 40 V/m barefoot on a wet cloth.

EM waves and pulses: Our experimenters had headaches during discharge experiments using a Van de Graaff generator. At their distance of about 1.5 - 2 m from the high voltage sphere, the pulse electric field intensity would have been about 10 V/m (0.25 W/m²). Experimenters who exposed rats to EM waves of 300 MHz at a power of 10 W/m² (electric field, 60 V/m)—the US regulatory limit for microwave power—remarked they felt depressed. The experimenters' station was two to three meters from the generation site, at which distance the power level should be less than 1 W/m² (20 V/m)—though the scattering of EM waves by nearby objects and interference might have enhanced the fields locally.

Prickling sensations: There are no reports of people's hair standing on end before earthquakes. However, some prickling sensations and the erection of small hairs on the hands were retrospectively reported to have occurred before the Kobe Earthquake. Some also heard crackling sounds as if a crisp plastic bag were being handled. Within the range of individual variation, hair standing on end would need higher voltages, which points to a probable upper limit to an electric field present before large earthquakes.

These reported human sensations argue for the appearance of an electric field before the Kobe and Gujurat Earthquakes of an intensity well over 10 V/m.

(b) Is a dog's tongue a sensor?

A pup placed on a wet cloth (Figure 4.10), detected electric field pulses through its toes and examined the wet cloth with its tongue at 70 V/m (about 3.5 V from leg-to-leg). It barked at 90 V/m (4.5 V leg-to-leg) as if to warn an invisible enemy. At 180 V/m it avoided the wet cloth on the ground. An adult dog, detecting an electric field of less than 90 V/m, avoided the area (Huang *et al.*, 1997).

Dogs were exposed to EM pulses using a Van de Graaff generator discharging to a grounded metal sphere about 5 m away. It was difficult to tell if they responded to the

Figure 4.10 Electric field effects on dogs on a wet cloth. Dogs detected a weak electric field corresponding to 3.5 V with the paw and tongue, and barked (Huang *et al.*, 1997) at the Chinese Anhui Seismological Bureau.

sound of the discharges or to the EM waves. One of them showed interest and approached the Van de Graaff, but turned back at a distance of 2 m (8 V/m). They closed their eyes several times, yawned, left the area or never approached it. In another experiment discharges using a Wimshurst generator scared one dog. Dogs' eyes and tongues seem more sensitive to EM pulses than their paws.

(c) Cats leaving with owner's socks

A Japanese TV program series called *Amazing Animals* was broadcast in March and July, 2000, based on our laboratory experiments on animal responses to EM waves. To double-check our results the director of the TV program repeated the experiment in his own home. While his cat was asleep in a room he brought the Van de Graaff generator into the next room and generated EM pulses by arc discharges. Figure 4.14 (a) shows the sleeping cat. In (b) she suddenly becomes aware of something, presumably the EM pulses of less than 8 V/m. In (c) her eyes close with each pulse. In (d) she leaves the room, meowing lugubriously, taking with her her owner's

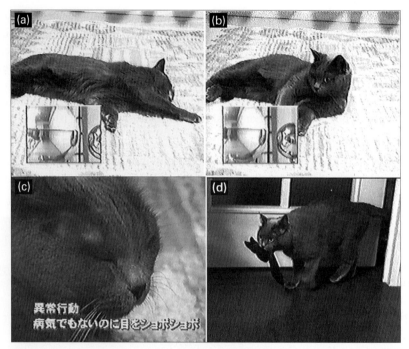

Figure 4.11 The sleeping cat in photograph (a) noticed the discharges in the next room as shown in (b), closed the eyes to pulses as in (c), then left the room meowing lugubriously carrying her owner's socks (d). (From a TBS TV program, *Amazing Animals*, broadcast on February 16, 2000.)

socks, carrying them as she would a kitten, though she had no kitten. (Animal psychologists say domestic animals will sometimes treat owners—and presumably owners' possessions—as one of their own, seeking to protect them.) The program attracted a good deal of interest: 20% of TV viewers on Saturday evening according to a survey.

In another program, electric discharges were produced in a room where several cats were present. The cats began to wash their faces at a distance of 4 m (a few V/m) and some showed gagging behavior. Apparently, their EM sensors are their paws and eyes, where sweat and tears having a high electric conductivity are secreted. Their sensitivity was higher than that for dogs.

A cat on a wet cloth flicked her tail and looked around suspiciously on application of an electric field of 45 V/m (leg-to-leg voltage of about 2 V), lower than that producing a response in a dog. She cleaned her paws and washed her face continually (Huang *et al.*, 1997).

(d) Horses sensitive to leaking electric currents from trams?

We had no opportunity to study the minimum electric field detectable by horses. However, many horses are reported to have received electric shocks when their hooves touched the rails of electrically operated trams, suggesting a high sensitivity to ground electric fields—presumably because horse shoes act as electrodes conducting the current flow to the body.

(e) Grooming Rats

Albino rats stood on a wet towel with a surface electric field of 2 V/m at DC voltage, or 20 mV for a leg-to-leg distance of 1 cm. They exhibited nervous and grooming behavior, as shown in Figure 4.12 (a). At 70 V/m leg cramps were observed in a sleeping rat (b), but the animals did not move violently. They groomed each other as in (c), and in (d), to avoid the ground current, supported their weight on their nails (rather than their paws) and moved round slowly and cautiously. No reaction was induced by discharges using a Wimshurst generator at a distance of 30 m (measured pulse electric field of 0.7 V/m), but they appeared startled at each EM pulse of 1.1 V/m at 15 m and body spasms began at 8 V/m at 2 m.

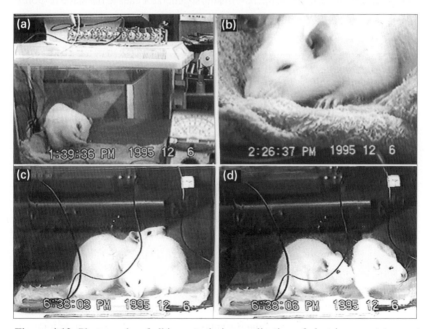

Figure 4.12 Photographs of albino rats during application of electric current to a wet floor. (a) Grooming due to application of electric current to the wet floor (F = 25 V/m), (b) leg cramps, (c) grooming each other and (d) a slow and cautious walk due to the increased field intensity (Ikeya *et al.* 1996b).

Small, black sand rats were placed on wet tissue on the floor of the cage. An electric field of 60 V/m was applied only to the right half of the floor by placing a copper electrode at the center. The rats quickly learned to move to the left where no voltage was being applied. When 100 V/m (1 V/cm) was applied to the whole floor, they stood up on their hind legs, ran, jumped up in panic and stood on the electrodes to avoid the electric field. The leg-to-leg voltage was only 2.5 V, which did not hurt them. Their actions were more dynamic than those of the albino rats.

(f) Guinea pigs

A guinea pig placed on a wet floor as in Figure 4.13 (a) got cramp in its legs at an electric field of 30 V/m. Its movement were nervous and it began crying when subjected to a field of 100 V/m, corresponding to a leg-to-leg voltage of about 5 V. At 400 V/m (a leg-to-leg voltage of 12 V) it began to groom itself (b) and stand on two legs using its claws (rather than its paws) to support itself against the glass (c) to minimize current flow.

An electric discharge between the high voltage sphere of the Van de Graaff generator and a grounded sphere at a distance of 3 m (8 V/m) from the guinea pig, produced cramps and made its hair stand up. A Wimshurst generator caused the same responses.

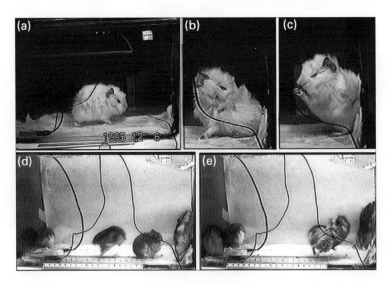

Figure 4.13 Guinea pigs: (a) before applying electric field pulses, (b) grooming at 30 V/m and (c) supporting itself against the glass. Hamsters: (d) avoiding current effects by standing on the right electrode, and (e) crying, running, and jumping in panic due to increased current.

(g) Hamsters: Sensitivity

Hamsters in the experimental site moved around constantly and so it was difficult to distinguish between normal and unusual behavior after the application of an electric field. However, they groomed desperately at a ground electric field of 30 V/m, and ran tumbling in panic at 50 V/m as shown in Figure 4.13 (d) and (e). Based on a leg-to-leg distance of about 3 cm, the voltage was 1.5 V at the most.

Hamsters exposed to EM pulses from a Van de Graaff generator showed very similar behavior, as shown in Figure 4.14, and in another experiment tried to hide. Their muscles convulsed with each EM pulse at a distance of 3 m (8 V/m). When arc discharges were applied using a Wimshurst generator, they responded to each pulse at a distance of 20 m (1.2 V/m, corresponding to a magnetic field of 4 nT and a power of 3.6 mW/m^2), as rats also did. They sneezed with each pulse at a distance of 30 m (0.8 V/m, 2.7 nT and 1.6 mW/m^2), though rats showed no response in this case.

The rodents were more sensitive to EM pulses traveling in the atmosphere than on the ground, meaning their sensors are their eyes, whiskers and tongues rather than their paws. However they might detect EM pulses directly through their nervous systems as discussed in the next chapter.

Figure 4.14 (a) The normal behavior of a hamster and (b) grooming and repeated washing of the face following generation of EM pulses by electric discharge from a Van de Graaff generator.

4.4.3 Birds: Puffed-up feathers, grooming, shrieking

(a) Red sparrows

Two red sparrows were placed on wet tissue paper on the floor of a cage, as shown Figure 4.15 (a). At a high field of 100 V/m (corresponding to only 1 V based on leg-to-leg distance of 1 cm) they puffed up their feathers (apparently to relieve discom-

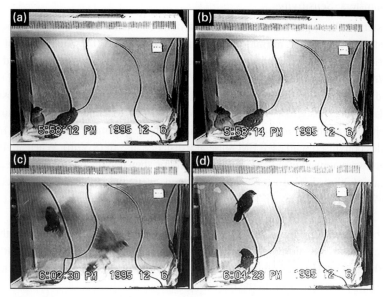

Figure 4.15 (a) Red sparrows on wet tissue paper (b) puffed up their feathers and bodies on the application of an electric field (c) flew about and finally (d) perched on the shielded wire to avoid the electric field.

fort), as in (b) and groomed themselves or sometimes each other, as the rats did. When the voltage increased, they first hopped, flew about as in (c) and then perched on the shielded wires to avoid the field as in (d) (Ikeya *et al.*, 1996b).

(b) High pitched twittering of yellow budgerigars

(i) Electric field pulse effects at 50 ms

This experiment was carried out at a Chinese high school using one yellow and one blue budgerigar. A perch in a cage was covered with wet clothes to which electric pulses were applied, as described in Section 4.4.1 (b). The birds shifted slightly with each pulse of 50 V/m, at about 130 V/m they puffed up their feathers and bodies, groomed and twittered in high pitched tones, and at 160 V/m, flew up to the side of the cage and stayed there. They stood on the perch on only one foot, presumably to minimize current flow to the body.

(ii) Air gap discharges

A Van de Graaff generator produced pulsed EM fields of 8 V/m at about 3 m from its sphere. A yellow budgerigar at 3 m distance twittered at high pitch with each discharge, in short trills as if alarmed. It was more sensitive to non-contact EM pulses than to pulsed electric voltage to the legs through the perch. It puffed up its feathers and moved its feet nervously.

(iii) Pulsed current to a loop antenna: EM pulses with no sound

Discharges generate sounds as well as EM pulses. To eliminate this possible interference, EM pulses were applied as shown in Figure 4.16 using a square antenna 30 cm in length, an amplifier and a pulse generator. At 0.5 V/m (EM pulses of 5 ms, frequency 5 Hz), the budgerigar inflated its feathers and stood on one leg on the perch. The yellow budgerigars left the perch for the metal cage at a voltage of 8 - 10 V/m.

Figure 4.16 A budgerigar exposed to EM pulses by placing the cage at the center of a loop antenna. The electric field of 0.5 V/m (as measured with a field meter) made it puff up its feathers.

(c) Pheasants and ducks: shrieking and alarmed

A pheasant shrieked at high pitch with each 5 V/m electric discharge from a Van de Graaff generator (at a distance of 3 m) and tried to get out of the cage. So, the legendary use of pheasants as earthquake early-warning agents could be due to their sensitivity to EM pulses rather than seismic P-waves.

Two ducks were placed on wet cloths to which electric pulses were applied using an "Elepulse" commercial electric massager. They each quacked at every pulse of 3 volts (the ground electric field was 30 V/m for a leg-to-leg distance of 10 cm).

4.4.4 Sensitive reptiles

(a) Snakes woken in winter by EM pulses

Snakes were placed on wet cloths between aluminium electrodes separated by about 70 cm. The experiment took place in winter. The snakes stayed torpid, but began to move when a pulsed field of 10 V/m was applied on both sides. Muscle cramping was observed corresponding to each pulse.

Electric discharges at about 1 m distance generated EM pulses with an electric field of about 16 V/m, at which level the snakes raised their heads to slide away from the acrylic cage, presumably seeking a more comfortable conductive surface.

(b) Lizards producing air bubbles

Red belly lizards produced air bubbles from the mouth in response to electric pulses at 15 V/m, presumably due to muscle convulsions. They moved quickly and climbed the aquarium wall in panic at 20 V/m.

4.4.5 Insects and worms

(a) Honey bees buzzing

An experiment on electric field effects on honeybees was carried out in the depths of winter on a mountain during the filming of *Amazing Animals*. The high voltage sphere of a Van de Graaff generator was placed close to a beehive with a transparent acrylic cover (Figure 4.17). Honeybees, whose movements are very sluggish in winter, were not agitated by the movement of an intense "Neomax" permanent magnet. They were, however, agitated by the electric field and swarmed in the direction of the high voltage sphere i.e. to the most intense part of the electric field. EM discharges to the edge of the box made them highly agitated and they came out of the hive and buzzed around loudly. The estimated intensity was 80-100 V/m at about 30 cm from the discharges.

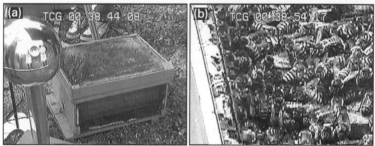

Figure 4.17 Honeybees exposed to pulsed EM waves generated by electric discharges from Van de Graaff and Wimshurst generators (From the TBS TV program *Amazing Animals*, 2000). (Video photo.)

(b) Stag beetles *(Dorcus hopei)* waking from hibernation

A plastic box containing two stag beetles sleeping in wood shavings in winter, was placed at a distance of 10 and 30 cm from the discharge of a Van de Graaff generator. The voltages of the pulsed EM waves were about 240 V/m and 80 V/m, respectively. The beetles woke up and moved actively (Figure 4.18), as they were reported to have done before the Kobe Earthquake.

(c) Ants: face-washing behavior

Ants, which have been reported to move to new habitats before earthquakes, were exposed to discharges using a Tesla coil. They began to wash their antennae and faces vigorously just as cats and rats did. Their antennae apparently act as sensors for EM waves. Assuming the antennae length to be a half wavelength, they might be more efficient at sensing EM waves at microwave range, and thus be able to detect lightning in approaching rain clouds, as cats can ahead of rainfall.

(d) Earthworms and lugworms emerged from soil or congregated

Aluminium foil electrodes were attached to opposite ends of a box containing wet soil. Worms emerged from the soil at 100 V/m when the voltage from a battery was switched on and off. They swarmed and congregated at the ends to avoid the electric field effect as they were observed to do before the Kobe Earthquake. (Uniform charge at the electrodes

Figure 4.18 Stag beetles awoke from winter sleep and emerged from the soil when an electric field and EM pulses were applied (From the TBS program *Amazing Animals,* 2000). (Video photo.)

minimizes the electric field.) When an AC voltage source was used, some small earthworms responded at a higher frequency than large ones in the same box. This is reasonable because when alternating current is used, the impedance or the current in the body of an animal will depend on the frequency and the size of the animal. Lugworms showed more pronounced movement.

(e) Silkworms in alignment

Ten silkworms at the fifth stage of growth, and their food beds, were kept in a box between electrodes. DC voltage was applied gradually up to 20 kV/m or suddenly with a stepped increase in voltage. They continued normal eating behavior, and only sporadically aligned themselves perpendicular to the electric field. This is

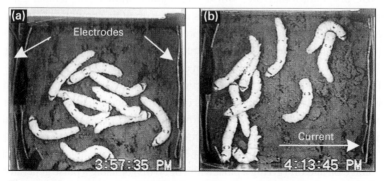

Figure 4.19 (a) Silkworms initially placed. (b) aligned perpendicular to the left to right direction of the electric field (Ikeya *et al.,* 1998b). (Video photo.)

natural since the polarization in such a conductive body quickly cancels the intense outer electric field. (It is not the electric field itself, but the current flowing in the body that causes unusual behavior.) When current was made to flow through the food bed they sensed the electric field and aligned themselves perpendicularly. The silkworms were initially placed as shown in Figure 4.19 (a), but they aligned perpendicular to the current as in (b) to minimize exposure to it. The observation that the distribution of silkworms was strange before the Kobe Earthquake may be due to their sensitivity to seismic EM waves. In their perpendicular alignment they were behaving like the minnows in Section 4.3.3.

4.4.6 Body current estimation

Experiments on electric field effects indicate that specific types of responses seem to depend on the particular species and on individual animals within the species. Unusual behavior began at low current densities for mice, rats and hamsters, yellow budgerigars, parrots and pheasants. The corresponding electric field intensities are shown in Table 4.1, 4.2 and 4.3 at the end of this chapter. The electric potential difference between the legs dictates the current causing the unusual behavior. A rough estimate of body current through the animals' legs can also be given using a measured leg-to-leg resistance of a few mega-ohm. The body current creating behavioral response is of the order of 0.001 mA, consistent with early work on catfish. In Galvani's experiments 0.002 mA made frog legs twitch.

4.5 Experiments at Kobe-Oji Zoo

4.5.1 Electric field effects on animals in the Kobe-Oji Zoo

(a) Californian sea lions: agitation and zigzag movements

The BBC program, *Tomorrow's World: Earthquake Animals*, September 23, 1998, reported that a couple of animal keepers at the Kobe-Oji Zoo noticed Californian sea lions acting strangely. The animal keeper said in the interview, "They were jumping around and looked agitated. I remember, we chatted with each other and said 'Maybe, this is a sign that there's an earthquake coming'".

The director of the Kobe-Oji Zoo participated in the Kansai Science Forum and allowed us to study electric field effects on sea lions and other animals at the zoo under the supervision of the keepers. Iron-mesh barbecue nets covered with an aluminium cooking foil were used as electrodes measuring 0.5 m x 0.7 m and dipped into a pool at separations from 1 m to 5 m. Electric fields were applied by switching DC voltage on and off (giving a pulsewidth of 50 ms) or using a pulse generator and an electric massager, Elepulse (Omron HV-FO4).

Figure 4.20 Response of a sea lion in a pool at the Kobe-Oji Zoo to an electric pulse of 0.5 V/m: (a) Swimming normally. (b) Noticing a weak electric field pulse having a peak intensity of about 0.5 V/m. (c) Turning quickly. (d) Avoiding the field (Ikeya *et al.*, 1997c). (Video photo.)

Electric field intensities at different sites were calculated using an electric field simulator and are given in Figure 4.5. A sea lion detected a single electric pulse of about 0.5 V/m, changed direction quickly and left the area as shown in Figure 4.20. It was highly agitated, and fussed and roared loudly on land. All the sea lions gathered on land and would not enter the pool.

(b) Hippopotamus: submerged and avoided the electric field

An electrode separation of 20 m was required due to the large size of the hippopotamus. This made it difficult to use a battery pack in the experiment to apply an intense electric field. The approximate field strength when the hippopotamus approached the electrode and submerged was a few V/m. The only recordable behavior was its leaving the site and sinking to the bottom of the pool.

(c) Green snakes: cramped and congregated

Green snakes were captured after they migrated into the Kobe-Oji Zoo after the Kobe Earthquake to eat birds' eggs. Several of these snakes were placed on a wet towel spread in a box. Both DC and pulsed electric fields were applied to the snakes, as shown in Figure 4.21. They began to spasm on application of the pulsed field and congregated to avoid it. The proverb,

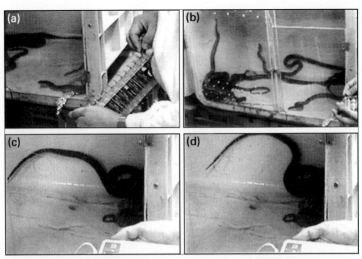

Figure 4.21 Green snakes (a) before and (b) after electric shock using batteries. In (c) and (d) the muscles cramped in synchrony with electric pulses produced by a commercial electric massager, Elepulse (Ikeya *et al.*, 1997c).

Snakes congregate in a bamboo bush before earthquakes,

may indicate that they use bamboo roots to shield themselves from EM pulses originating underground and congregate presumably to minimize the exposure effect, as seen in earthworms.

(d) Birds: avoid electric field

Flamingos and geese were subjected to electric fields from 4 to 15 V/m. One goose appeared to detect when the voltage was on or off and showed slight convulsions at 15 V/m.

Penguins avoided the pool when pulsed electric voltage was applied to the electrodes and gathered in a nearby pool instead.

4.5.2 Estimated electric field intensity before the Kobe Earthquake

Reports of fish alignment and paralyzed fish before the Kobe Earthquake might indicate electric field intensities of more than 10 V/m in water. The responses of alligators are similar to those of minnows but crocodiles and sea lions are at least one or two orders of magnitude more sensitive than other animals.

The reported fish alignment direction before the earthquake, at Nishinomiya city and at Takarazuka (respectively 25 km and 35 km east of the epicenter) was almost east-west. Because fish align themselves perpendicular to electric fields to minimize discomfort, the electric field probably ran north-south, perpendicular to the propagation direction of EM waves which must therefore have emanated from the west (or from the east), as explained in Section 3.3.2. The field intensities from 1 V/m to about 50 V/m and their directions are estimated from retrospectively reported animal behavior and indicated on the map (Figure 4.22) together with the numbers of reports and their distribution before the earthquake.

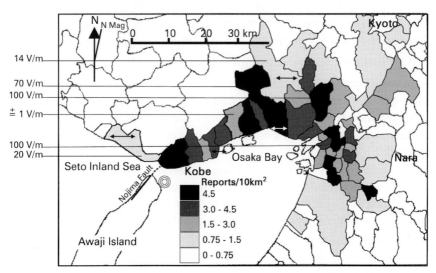

Figure 4.22 Estimated intensities of preseismic electric fields at several places based on reported unusual animal behavior before the Kobe Earthquake. Numbers of reports per 10 km² are indicated by shading in the map, together with the fault (line) and epicenter (double circles). There were 5.3 reports per 10,000 residents close to the epicenter area. The arrows represent the east-west alignment of floating fish, perpendicular to the electric field direction but parallel to the direction of EM waves (Ikeya *et al.*, 1997c).

The experiments described in this chapter suggest that if animal reactions observed in the Seto Inland Sea before the Kobe Earthquake were authentic, an electric field of more than 10 V/m in water must have been present. This corresponds to an electric field of 90 V/m of EM waves in air, and a peak power of about 20 W/m². The total EM energy would then be about six billion watts at the source, based on a radius of about 10 km from the epicenter, or about 100,000 joule for a pulsewidth of 0.1 ms. If six billion watts is generated by a rock volume of 1 km³ [the volume of the Kobe Earthquake was 1000 km³, Section 3.2.1)], the EM energy is still only 0.1 J/m³. Six billion watts is about one ten thousandth of the total energy released by the main shock and, therefore, a physically reasonable level of energy to be released in the form of EM waves by a large earthquake.

4.6 Izu-Atagawa Tropical & Alligator Garden

4.6.1 Crocodiles: Eyes as electric sensors

Crocodiles at the Izu-Atagawa Tropical & Alligator Garden in Japan growled and behaved violently before volcanic eruptions at Izu-Ohshima in 1971 and before several earthquakes in the Izu area. Alligators and crocodiles have responded similarly a day before approaching thunderstorms. The reports suggest they were responding to pulsed EM waves created by lightning from approaching thunderclouds, just as cats proverbially wash their faces before bad weather.

(a) Electric field effect: sensitive crocodiles

Alligators (*Family Alligatoridae*); Spectacled caiman (*Caiman crocodilus* Linne): This species is sensitive to climate changes but has not shown distinctively unusual behavior before earthquakes. Younger members of the species showed a startled movement in response to an electric field of 4 V/m (like catfish) and bit the electrode.

    Crocodile (*Family Crocodylidae*): Paraguay caiman (*Caiman crocodilus* Yacare): This species responded to an on-an-off switching of a 5 V/m DC voltage by closing and opening its eyes at every switch [See Figure 4.23 (a) and (b)]. Cramping round the mouth started at 6 V/m. At 9 V/m the crocodiles became excited and violently bit the electrodes, at about 10 V/m they experienced body convulsions. Their responses to pulsed electric fields started as low as 0.02 V/m for one or two sensitive crocodiles in a group of 12; at 0.5 V/m all were responding. So it could well be that the crocodiles at the Darica-Kocaeli Bosphorus Zoo, did refuse to enter their pools—a conductive medium for EM waves—before the Izmit Earthquake.

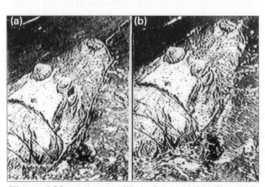

Figure 4.23 (a) A crocodile before and (b) after application of an electric pulse. It closed its eyes with every pulse. (Video photo.)

(b) EM pulses by discharges: closing eyes at each pulse

Discharges from a Van de Graaff generator:

When arc discharges to the ground from a high voltage sphere of a Van de Graaff generator were made about 1 m away from the crocodiles, they closed their eyes at each pulse. Because no intense light is produced by the discharge they apparently detected the generated EM pulses with their eyes. Crocodiles maintain their salt balance by releasing urine and tears. Conductive eyes and connected nervous systems might act as an antenna for EM waves.

Low body resistance:

Body resistance measured at the back is more than 40 megohm (MΩ) for baby alligators and crocodiles, and exceptionally small (0.5 - 1 megohm) in the belly skins of some crocodiles. Different current flow would be a main cause of sensitivity in some crocodiles. Pressing their bellies to the wet ground, they may be able to sense electric signals from prey animals on land.

4.6.2 Turtles (eyes) and snakes (tongue)

(a) Turtles (*Hieremis annandali* Boulenger) closing eyes and washing faces

Turtles from Thailand retracted their heads at about 2 V/m and closed their eyes with each electric pulse of 4 V/m. One turtle attempted to wash its face like a cat

(See Figure 4.24). Turtles also physiologically balance their salty body fluids in the same way as cats and crocodiles. Their skin has no sweat glands, so they will not detect electric fields in the skin. However, their conductive eyes with a low resistivity may act as electrosensory organs able to detect electric fields.

(b) Snake (Boa Constrictor)

When electric pulses were suddenly applied to a tree in which a boa constrictor lay, it moved its tail and tongue. A wet tongue with sensitive innervation might be an antenna for electric fields produced by EM waves. The snake also adopted an S-configuration in its upper body: a pre-attack stance.

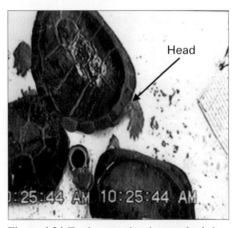

Figure 4.24 Turtles experiencing a pulsed electric field. They closed their eyes with each pulse and "washed their faces". (Video photo, Izu-Atagawa Banana Alligator Garden.)

4.7 Animal experiments in Turkey and Taiwan

4.7.1 Parrots at the Darica-Kocaeli Bosphorus Zoo

A newspaper in Ankara, Turkey, reported that parrots at the Darica-Kocaeli Bosphorus Zoo were twittering in high tones before the Izmit Earthquake. An ion spray gun using a piezoelectric element (capable of causing EM noise) was used to check whether parrots could detect the weak EM waves that cause noise in a radio receiver. They appeared to be avoiding the weak electrostatic field of less than 1 V/m at 10 cm, though sunflower seeds were offered as an inducement.

4.7.2 Earthworms at an elementary school in Kaoshung, Taiwan

Kaoshung is a southern city located along the extension of the Chehlungpu fault line. Earthworm swarms were reported 16 and 8 days before the big earthquake (M7.2) on November 2, 1999, six weeks after the Taiwan-921 Earthquake (M7.7), (Figure 2.11).

An on-and-off voltage was applied using a battery pack at the place where thousands of earthworms had appeared. A few earthworms appeared on the surface when the apparent electric field increased to about 100 V/m (Figure 4.25). However, when we excavated the soils, many had swarmed to lower levels where the electric field was much less intense due to the form of the electrode (See Figure 4.5). If an electric field had been present before the earthquake, earthworms would have emerged at a much lower voltage, though the approximate intensity of the electric field would still have been around 10 V/m judging from the reports. At that level unusual behavior may also have been occurring in other animals.

Figure 4.25 (a) Experiments on electric field effects on earthworms at the Kaoshung Elementary School, where thousands of earthworms appeared eight days before a large earthquake on November 2, 1999. (b) Earthworms having emerged after application of an apparent 100 V/m.

4.8 Preseismic electric field intensity at epicenters

4.8.1 Electric pulse effects on aquatic animals: Galvani's frog muscle

Extrapolating from this chapter's experiments on electric field effects, unusual animal behavior before earthquakes could be explained as behavioral responses to electrical pulse stimuli (an electric field of EM waves).

Local stress changes in rocks generate charges, presumably by releasing piezo-compensating charges during microfracture or by some other mechanism—like frictional electricity or fluid flow electrokinetics (Section 3.3.8)—and any generated electric dipoles, as they collapse, radiate EM waves as EM pulses. The low frequency component with large skin depth comes out close to the ground inducing charges on the surface. Induced current from bedrock granite with a resistivity of 1 -100 k ohm-m (kΩm) in conductive wet sediments (100 Ωm), in the sea (0.01 Ωm), and in rivers (50 Ωm) would explain the unusual behavior of aquatic animals before earthquakes as an electrophysiological response.

Aquatic animals have extraordinarily sensitive electro-receptors for communication and capturing prey animals. Some fish without electrosensory organs are thousands or a million times less sensitive to electric fields, but have also been reported to have exhibited unusual behavior before earthquakes. The electric field intensity, leading to cramping in ordinary aquatic animals, is about 10 V/m, corresponding to a current of a few microampere, which was sufficient to produce convulsions in Galvani's frog muscles. A field of at least this intensity would appear before large quakes.

4.8.2 Land animals may still respond to electric pulses

Land animals may have electric sensors not yet appreciated by scientists. The sensory organs that perceive pulsed EM waves would be the conductive parts of the body connected to nerves and synapses; moist eyes and paws and skin containing sweat. Experiments in this chapter show that it is the body current rather than the electric field itself which affects animals. An electric field intensity of 10 - 100 V/m in air was sufficient to cause muscle convulsions in frogs' legs and would also cause unusual grooming behavior.

These electric stimuli, inducing current in some animal bodies, disrupt the equilibrium sufficiently that animals can become disturbed, anxious, irritable, panicky, aggressive towards each other, and exhausted. The next chapter shows that these electric stimuli affect production of neurotransmitters in the brain.

Note that the mobile charges (electrons and holes) and the high frequency component of EM waves cannot travel from the underground fractured crust to the surface, so cannot cause unusual behavior in animals.

4.8.3 Are intense EM pulses observable before earthquakes?

The probable presence of electric fields of 10-100 V/m (0.25 W/m^2) at the epicenters of large earthquakes has been estimated from the experimental responses of ordinary animals to induced body currents and from reports of unusual animal behavior. Geophysicists and seismologists have observed earth voltages of only 0.05

mV/m before earthquakes, but these voltages are far too small to cause animal behaviors of the kind we have been describing. However, animals behaving in unusual ways, both experimentally and at earthquake sites, are responding to a peak intensity of EM pulses in a wide range of frequencies, while geophysicists observe only DC and the intensity of a narrow band (1 Hz) in a particular frequency range. If only the narrow frequency distribution of a pulse is considered, the intensity in the frequency range close to DC (0 Hz) may very well be a million times lower than the peak value we have been discussing. In addition, if the specific dielectric constant of ground is about 10 - 100 million at ULF (due to surface water on rock minerals), the electric field intensity measured by geophysicists and seismologists would indeed be orders of magnitude smaller than that of EM pulses (See Section 11.3.8 for a fuller explanation). So there is not necessarily any inconsistency.

Readers may wonder why animals are not continually behaving in odd ways in modern cities full of EM waves produced by TV and radio broadcasts, home electric appliances, and especially modern mobile phones. This is because the EM electric fields are much less intense than 1 V/m and are mostly continuous waves, not pulses. The only exception is a mobile phone with a power of about 1 W at the source, (about 1.2 V/m at a distance of 3 m from the unit). In addition, our experiments show that animals respond only to pulsewidths of more than 0.1 ms (one ten thousandths of a second) and most animals in cities would be receiving pulsewidths of less than 1 μs (one millionth of a second). They will adjust to them in a modern city environment of artificial EM noises, only behaving in unusual ways when stimulated by natural EM pulses (of 1-100 W/m²: 20-200 V/m: 60-600 nT) induced by lightning and before earthquakes (Magnitude > M4; distance from epicenter < 30 km, pulsewidth > 0.1 ms, intensity > 1 V/m).

People using mobile phones at a distance of about 5 cm from the head, generate a power density of roughly 25 W/m² and so a field of about 100 V/m. What makes the effect difficult to estimate is the scattering factor and induced current which depend on the frequency and body resistivity (impedance), as described in Figure 3.17. The author knows several persons who complain of headaches when they use mobile phones and others who have not been able to sleep on the night of local early morning earthquakes of moderate magnitude (M5). As mentioned, experimenters had headaches using the Van de Graaff and Wimshurst generators.

There have been no reports of people killed in an epicenter area by electric shock from precursory or coseismic EM pulses, though there have been reports of fish floating dead before earthquakes. This argues for a maximum electric field intensity of less than 10 - 20 V/m in water (90 - 108 V/m in air) for an M7 earthquake at a distance of less than 100 km from the epicenter.

To reiterate, a clear distinction needs to be made between magnetic and electric fields in discussing unusual animal behavior. Because a rapid change in a magnetic field generates electric fields, it is easy, though incorrect, to attribute the behavior to the magnetic field. The magnetic field itself, at its weak intensity, does not create biological effects; it is the electric field that induces current in the body depending on its impedance.

4.9 Summary

Unusual animal behavior before earthquakes would be caused by pulsed electric fields rather than magnetic fields whose intensity changes before earthquakes are reported to be less than one ten thousandth of the Earth's magnetic field. (Converted to EM waves such a change would correspond to a few V/m.) Experimental conditions precluded the generation of EM waves in every experiment we conducted, but animals that responded to electric field experiments will *ipso facto* respond to EM waves of less than 10 kHz. Behavioral responses of both sensitive and ordinary animals to electric fields of EM pulses are summarized in Table 4.1. Effects of electric field intensity on some animals are summarized in Tables 4.2 and 4.3, respectively.

Based on the reported unusual behaviors in animals before large earthquakes, and their laboratory responses, the author considers that intense pulses with a duration of more than 0.1 ms, and an electric field component of about 20 V/m (1 W/m², 60 nT) or more, must appear at such epicenter areas. The much smaller electric fields and voltages measured by geophysicists and seismologists in units of [mV/km] (one millionth of the intensities we have been discussing in units of [V/m]), can be accounted for by their measurement of DC in narrow 1 Hz bands, and the magnitude of the dielectric constant (discussed in greater detail in Chapter 11).

Table 4.1 Animal behavior at an electric field intensity with a pulsewidth of more than 0.1 ms.

Animals	Behavior	Electric field	Power*	Magnetic field*
Sensitive	panic	2 - 3V/m	0.01 W/m²	10 nT
Ordinary	muscle cramps	10 - 20 V/m	1 W/m²	30 - 60 nT
(c.f.) Large quake (M>7)		60V/m	10 W/m²	200 nT

* Indicates the corresponding power density and magnetic field for EM waves.

Table 4.2 Effects of electric field, F (V/m) on fish,worms, other animals and insects for estimating the critical current density, J (A/m²) to explain unusual animal behavior before earthquakes.

Animal	F (V/m)	Response
Catfish	3-4	Violent movement
Eels	0.2	Violent movement
Goldfish	15	Startled reaction to initial pulse
	23	Oriented body perpendicular to field
Minnows	10-15	Q: Quick movements with on-and-off voltage
	25	A: Aligned perpendicular to the direction of an electric field
	70	P-P: Panic, some paralyzed
Guppies	10	Q
Loaches	4	A
	30	Q
	70	P-P
Earthworms	100	Came out of soil and swarmed
Lugworms	100	Came out of soil and swarmed
Tortoises	2	Movement
	5	Walked to edges
	30	Tried to climb the wall
	50-75	Climbed on others
	100	Ran rapidly in panic, neck cramps with voltage on & off
Frogs	1	Breathed fast
	4	Slight movement
	10	Washed face, moved to edges
	12	Jumped and violently moved
	24	Jumped to the edge
Lizards	15	Responded with mouth-bubbles to "on"
		Muscle cramps with pulses
	20	Moved quickly
	28	Climbed the wall in panic
Bees	50	Became agitated and came out of beehive

Table 4.3 Effects of electric field, F (V/m) on rodents and birds in explanation of unusual animal behavior. The voltage was applied to electrodes separated by 25 or 30 cm on the floor of the cages, between which wet tissue papers with resistivity of 20 kΩ were placed. The voltage between legs, V_{1-1} with a separation, L_{1-1} (cm) and the measured leg-to-leg resistivity, R_{1-1} are shown with the current, I_{1-1} (Ikeya *et al.*, 1996).

Animals	F (V/m)	V_{1-1}(V)a)	I_{1-1} (mA)[a)]	Responses [b)]
Human finger	25			Stimulated weakly
body	8			Sensing with body hair
Electric Health bath*	20			Muscle relaxation
Rats	2	~0.08	~0.15	Grooming (G)
(Rattus norvegicus)	6	~0.3	~0.5	G, Nervous looking (N)
W = ~300 g	72	~5.0	~6.0	Cramped legs (CL)
L_{1-1} = ~3.5 cm	600	~15	~52	Avoiding field (AF)
R_{1-1} = 0.4 ~0.5 MΩ	1000	~24	~90	Running (R), Panic (P)
Mongolian gerbils	60	~1.5	~0.75	G, Crying?, AF?
(Meriones unguiculatus)	100	~2.5	~1.3	Standing up (SU), G, N
W = ~50 g, L_{1-1}= ~2.5 cm	240	~6	~3	CL, AF
R_{1-1} = ~2 MΩ	400	~10	~5	R, P, Screaming (S)
Djungarian hamsters	30	~0.6	~0.3	Biting wires, A?
(Shangarian hamster)	50	~1.0	~0.5	Running in panic?
R_{1-1} = ~2 MΩ	400	~8	~4	G, Jumping (J)
W = 20 g, Ll-l = ~2 cm	800	~16	~8	R, P, S, Tumbling (T)
Guinea pigs W = 106g	100	~3	~0.15	Nervous demeanor?
(Cavia porcellus)	400	~12	~0.6	SU
L_{1-1} = ~3 cm	800	~25	~12	G
R_{1-1} = ~2 MΩ	1600	~50	~25	P, J, T
Birds				
Red Avadavat	100	~1	~0.5	Puffing up, Grooming
(Amandava amandava)	300	~19	2~3	Jumping, AF
R_{1-1}= 2 ~3 MΩ	660	~50	~5	Flying up (FU), P, A

a) The V_{1-1} and I_{1-1} were calculated as maximum to animals parallel to the field direction.

b) Behaviors are abbreviated as indicated and some that could not be judged clearly as an electric field effect are indicated by ? mark.

* Some elderly people enjoy electric baths as a health treatment, with electrodes producing voltages up to 20 V/m, the maximum permitted under Japanese Government regulations.

5

Unusual Animal Behavior: II

Rock Compression and
Increased Animal Activity

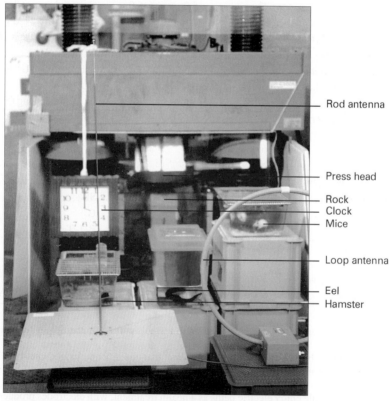

Experiments on compression and fracture of rocks 15 cm x 15 cm x 30 cm, to simulate an earthquake of magnitude M -2, were undertaken to generate EM waves and to observe simultaneous animal behavior. (Enlargement.)

5.1 Introduction

Reports of unusual animal behavior are always open to question. What is the measure of unusual behavior? How much is the respondent inventing or exaggerating the behavior after the fact? These are legitimate questions and on that basis some scientists reject precursor reports based on animal behavior.

This chapter attempts to objectify animal behavior. An experimental environment was set up to study animal responses to EM waves and rock fracture (simulating a tiny earthquake). Resulting changes in the levels of neurotransmitters in the blood and brain fluids of the exposed animals were measured in collaboration with veterinary physiologists (Ikeya *et al.*, 2000).

Disturbed day and night activity levels in mice before the Kobe Earthquake are presented as the first published, objective scientific evidence of unusual animal behavior before earthquakes (Yokoi *et al.*, 2003a). The heart as a possible mechanism of behavior change was also investigated for physiological response to EM waves. Two other small experiments testing for responses to EM waves are described: one at a chicken farm and the other a small controlled study involving mice. The experiments are attempts at scientific investigation and are offered for consideration. Suggestions are made about ways in which animal behavior might be an attempt to minimize or avoid EM fields, or to find electric shields. Because a causal pathway between EM pulses and brain response has not been established, the chapter also speculates about possible mechanisms by which animals might detect and respond to EM waves e.g. the structure of the electrosensory organs of catfish, and the possibility of muscle as a detector. A brief hypothesis about the action of melatonin is acknowledged to be highly speculative and obviously awaits clarification by specialists in these fields.

5.2 Rock compression experiment

Before main rock fractures causing large earthquakes, there are smaller rock fractures. Preceding these, microcracks may be formed by intensifying stress on the rock.

A laboratory experiment in which rock is compressed should thus reproduce the preliminary stage of an earthquake. If EM pulses generated by microfractures in rocks are responsible for precursor phenomena, then compression of rocks should also cause unusual animal behavior (Ikeya *et al.*, 2000). Hence, changes in neurotransmitters in brains and blood of animals placed near rock under increasing stress in a laboratory are good indicators that rock compression produces changes in animal behavior, probably by the production of EM waves. Such generation of EM pulses would need to be confirmed during this process of rock fracture.

Little work of this nature appears to have been done since a small rock of gabbro (a variable, coarse-grained basalt rock) was compressed near mice and their behavior recorded by Gawthrope *et al.*, (1978), as reported in a preliminary work in the Proceedings of the International Conference on Abnormal Animal Behavior Prior to Earthquakes. Mice showed fear by freezing their postures just ahead of the rock fracture.

Our experiment was undertaken in collaboration with veterinary physiologists, biologists, construction engineers and physicists to observe EM waves, acoustic emissions (AE) and the simultaneous behavior of several animals during the compression and fracture of about 25 rocks using 200-ton and 500-ton compression machines. Chemical substances in rats' blood were analyzed before and after rock compression and exposure to EM waves.

(a) Experimental conditions

Rocks of granite, marble and basalt, 15 x 15 x 30 cm^3 in size, rigged with strain gauges and piezoelectric elements to measure acoustic emission (AE), were fractured using mostly a 500-ton press, but also 200-ton and 1000-ton presses. The rock size corresponds to an M -2 earthquake, equivalent to typical very small and frequent, though undetectable, underground earthquakes—about one-billionth the magnitude of an M4 earthquake. An M4 earthquake may startle people at the epicenter but will not cause damage.

Two budgerigars, six rats and six mice were placed in cages, two eels in aquaria and eleven silkworms in containers, 1 m from the rock. They had adjusted to their environment before the compression began and were inert at the time. Their behavior was recorded by video camera. Researchers stayed at a distance so as not to disturb the animals. The experimental arrangements are shown in Figure 5.1 and in the picture on the title page of this chapter.

All detectors of acoustic emissions (AE), EM waves, test animals, sample rock and video camera were placed inside a 1 mm thick aluminum shield surrounding the press. The detection of pulsed EM waves was made using a one-turn loop antenna (228 MHz), wide band amplifier, active rod antenna, booster, electric field-meter (30 MHz, Bandwidth 9 kHz), and digital storage oscilloscope (Bandwidth 400 MHz). Another set of a loop antenna, a storage oscilloscope and monitoring TV were placed outside the shield to observe background EM noise and animal behavior.

Figure 5.1 (a) The 1000-ton compression machine used for the rock crushing experiments (Research Laboratory of Civil Engineering, Kohnoike Co.). (b) Layout showing experimental materials. (c) A broken rock specimen. (Video photos.)

(b) Stress, strain, EM waves and acoustic emission

(i) Stress:

Neither EM pulses nor any distinct differences in animal behavior were observed over a 15 minute period between rock compression and fracture at a pressure of about 330 tons (a stress of about 160 million Pascal [MPa]). A slow rate of stress increase of less than 5 tons/min (2.5 MPa/min) was necessary to generate EM pulses to the antennae and to produce unusual animal behavior. At that rate it usually took more than one hour from the start of compression to fracture.

Granite rock specimen

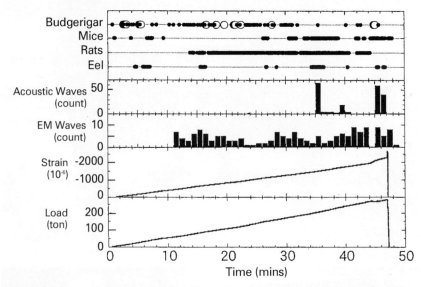

Figure 5.2 The variation in stress (load in tons) and strain for granite using the 500-ton press, together with the observation of EM and AE pulses over time. Animal activity is indicated at the top of the figure (See Section 5.2.2 for comment).

Figure 5.2 shows the observed stress, strain, acoustic emission (AE) and EM emissions as a function of time after the start of compression. The rock strain increased linearly as the pressure increased. Both EM and AE pulses were generated by the compression, presumably due to microfracture within the rock.

(ii) EM Pulses:

Pulses of EM waves with a width of about 30 nanoseconds and intensity of 10 - 50 mV/m were observed at frequencies higher than 100 MHz after 10 minutes from the start of compression at about 60 tons (30 MPa). Calibrating the efficiency of the antenna, amplifier and other measuring devices indicated a maximum EM pulse intensity of about 2 V/m (0.01 W/m²) at a distance of 1 m. The background level, lower than 30 MHz, showed no change during the experiment.

(iii) Acoustic emissions (AE):

These were detected at 35 min onwards, much later than the EM waves. The detection of EM waves and changes in animal behavior occurred well before the fracture.

(c) Fractures of various rocks:

(i) Pre-heated and wet granite:

There were no clear changes in animal behavior or marked detection of EM waves after granite rock samples were preheated to between 200 - 500° C. This may be because local microfractures occurred from internal stresses during the heating and cooling processes, leading to fewer fractures and fewer EM pulses during compression.

Less intense EM pulses were generated for wet rocks (immersed in water for 12 hours before the experiment) than for dry granite. If a fluid diffusion model is considered as a generator of EM pulses [Section 3.2.4 and Section 11.3.9 (b)] sufficient fluid in the rock increases conductivity and current flow reduces voltage buildup. So immersing the granite in water increases electrical conductivity and leads to quick recombination of piezo-compensating charges of opposite signs, both reducing the electric field intensity under stress and the generation of EM waves.

(ii) Marble and basalt:

Compared with granite, less intense EM pulses and little unusual behavior were observed during the fracture of basalt and marble. A tiny quantity of quartz dust is found in marble when it is dissolved in hydrochloric acid. Theoretically 1% of quartz dust in a sample produces 10% piezoelectricity, as if 10% of quartz dust is present. This arises from the strong piezoelectric effects of multiple small grains, even at random orientation (Chapter 3, Section 3.3.7). This suggests that piezo-compensating charges around the quartz grains may be responsible for EM emissions during compression, though other processes, which are an order of magnitude less effective than piezoelectricity, must be considered for basalt which has little quartz.

5.2.1 Analysis of blood plasma and brain hypothalamus fluid in rats

Hormones of rats placed near the press were analyzed to see if the crushing of rock, as in earthquakes, and/or EM waves could create physiological effects in animals.

The following neurotransmitters, among many hormones, are involved in the transmission of nerve impulses: adrenaline, noradrenaline, serotonin, dopamine and GABA or gamma-aminobutyric acid (See Table 5.1 for their properties). They are important to study because they can be involved in fear reactions that could connect to the animal behavior observed. They can be manufactured at various places in the body, particularly the hypothalamus, a small part of the brain thought to be directly involved with action in animals (particularly fear) and the sexual instinct, and are classified as part of the body's sympathetic nervous system.

These neurotransmitters were measured during and after the rock crushing experiments and EM generation in two ways by Professor Ohta's group in the Department of Veterinary Physiology, Osaka Prefecture University. First by withdrawal of small amounts of blood from conscious rats, and secondly, by passing a special artificial cerebrospinal fluid through the hypothalamus region in the brain, and analyzing it, after collection, by a technique called high precision liquid chromatography.

Table 5.1 Changes in neurotransmitters and hormones in blood plasma and hypothalamic fluid induced by rock compression (Com.), exposure to EM waves and charged ions (aerosol).

Chemical — Function	EM	Com.	Ions+	Ions-
Adrenaline: A hormone secreted by the medulla of the adrenal gland of the kidney by stimulation of the sympathetic nervous system. An index of uneasiness preparing body for fight or flight reaction; it increases cardiac frequency, blood pressure, dilates the blood vessels of the heart and brain and causes erection of hair.	Dec.	Dec.	Dec.	Dec.
Noradrenaline: A transmitter substance produced at the ending of adrenergic nerves giving effects similar to those of adrenaline. An indicator of excitement.	Inc.	Inc.	Dec.	Dec.
Serotonin: A pharmacologically activating compound, derived from tryptophan (one of the 20 essential amino acids). It expands the blood vessel by movement of muscle mainly controlled by the sympathetic nervous system, increases the capillary permeability and contracts smooth muscles.	NC	NC	Dec.	NC
Dopamine : Decarboxylation product of DOPA (dihydroxyl phenylalanine), a precursor leading to melatonin formation.	Dec	-		
Gamma-aminobutyric acid (GABA) $NH_2(CH_2)_4 COOH$ An amino acid which has an amino group on the third carbon away from the carboxyl group and acts to inhibit the transmission of nerve impulses. An inhibitor of noradrenaline.	Flu.	Flu.	Inc.	Dec.

* Decrease (Dec.), Increase (Inc.), No change (NC), Fluctuation (Flu.).

Both blood and cerebrospinal analysis gave similar results, and showed hormone levels moving consistently, but the direction varied with the hormone. The significance of the change of level of the hormones was checked using a statistical technique called multifactor analysis (Ohtani *et al.*, 1997). The results were taken as showing a significant change if the probability that the conclusion is wrong was less than 5% (often expressed as $p < 0.05$).

The results for two hormones are shown in Figure 5.3 and indicate significant changes for noradrenaline and adrenaline—noradrenaline increased and adrenaline decreased (Ohta *et al.*, 2000). This occurred both for granite fracture and exposure to EM waves. A decrease also occurred for dopamine, but GABA fluctuated, and serotonin did not change. These experiments showed for the first time stress reactions to the crushing of rock and to EM waves that were quite new to the physiological research community.

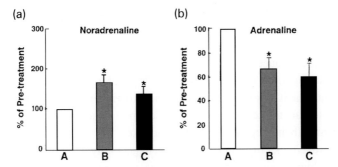

Figure 5.3 Plasma level of (a) noradrenaline and (b) adrenaline in the blood plasma of rats (A) before and (B) after exposure to EM waves at 300 MHz and 10 W/m^2 and (C) to granite fracture using a 500 ton press. The means and standard errors of 3-4 rats are given and the significance level (p <0.05) by asterisks (Ikeya., 2000).

There are therefore definite effects on the neurotransmitters in animal bodies but detailed interpretation is not completely clear, since adrenaline decreased (which should normally mean less fear) but noradrenaline increased, which generally means more excitement. More usually they would rise together.

To complicate this interpretation, levels of noradrenaline and adrenaline in the blood and urine of dogs fluctuated, but in an unclear way when exposed to EM waves with electric fields of 20 V/m at 200 MHz for both cw- and modulated waves (Aikyoshi, private communication, 2003). Rock crushing experiments were not used.

Ohta *et al.,* (in preparation) have found that mice exposed to EM waves have tended to form two groups in their adrenaline response. A decrease in adrenaline was only found in a minority of mice (between 20% and 30%) and in the rest adrenaline production remained more or less stable. In repeated experiments the presence of the two groups has persisted. Though the reasons for this are not clear there are implications that the minority group was closer genetically to mice with greater susceptibility to electric shock (See Section 5.3).

Wadatsumi and Ohta, (1998) have thought that charged aerosols may be involved in the reactions of animals to earthquake precursors (Tributsch, 1982), and have exposed rats to heat and cold, and in other experiments, positive and negative ions generated by appropriate pieces of equipment, analyzing most of the same neurotransmitters in blood and hypothalamic fluid. They also observed definite changes, some of which lasted longer than 30 minutes, but the directions of change depended on whether the ions were positive or negative, and the ion exposure results differed from the results of heat and cold exposure, and also from the EM wave/rock crushing experiments.

We are not likely to get real results on neurotransmitter levels in an earthquake, which would enable us to decide what the causative factors are, but there are real effects and they can be caused either by EM waves, or rock crushing. Given that the EM electric field produced in the rock crushing experiment was about one thousandth of what might be generated by EM pulses before a large earthquake we could speculate that it may have been insufficient to trigger adrenaline production, but sufficient to start an excitement reaction. Certainly some of the animal behavior was consistent with that reported before earthquakes.

The EM wave/rock crushing results do seem generally to be inconsistent with the Tributsch hypothesis that serotonin would increase in animals following current flow in the body after exposure to aerosols, because no change was observed. In fact the anxiety states observed in our experimental animals exposed to electric pulses would be consistent with a decrease in serotonin, not the increase suggested by Tributsch.

5.2.2 How did the animals behave?

Mice and rats: Figure 5.2 also shows the behavior of several animals as a function of compression time. The sleeping rats woke up, stood up, looked unusually alert, and groomed restlessly when the load on the rock was above 60 tons—one fifth of the load leading to rock fracture. They also washed their faces nervously as they did in a pulsed field of 2 V/m applied to a wet floor [Section 4.4.2 (e), Ikeya *et al.*, 1997b], and during exposure to EM waves of 1 W/m^2 at 300 MHz. They froze just before and during the fracture. Mice groomed nervously before the detection of EM waves at the antenna and froze when EM pulses were detected. They began grooming again and then bit each other in excitement. Chickens, hamsters and pigs are reported to have attacked or bitten each other before earthquakes.

Budgerigars: These puffed themselves up and twittered, or fluttered out of the perch area, as indicated by dots and open circles respectively. The detection of EM waves accompanied the behavior well before the fracture at about 300 tons pressure.

Silkworms: Specialist observation would have been necessary to determine whether apparent body alignment really occurred. An unusual lifting of the head was also observed.

Eels: Eels moved early in the compression process. They moved restlessly and sometimes appeared to have been electrically shocked when EM waves at the antennas were detected in the oscilloscope (Figure 5.4). The observed pulses (about 2 V/m) were intense enough to cause agitation in sea lions and crocodiles, but not for ordinary fish to align themselves to the field or worms to swarm and congregate (Chapter 4).

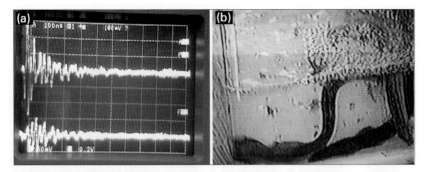

Figure 5.4   (a) Oscilloscope display of EM waves detected by loop and rod antennae, (b) simultaneous eel activity (Ikeya *et al.*, 2000). (Video photos.)

Human beings: Obviously psychological or unrelated physical factors may be responsible, but some of us experienced pain at the back of the head and fatigue after the fracture experiments—the same type of fatigue experienced during electric discharge experiments using a Van de Graaff generator and EM exposure using a 200-300 MHz antenna.

5.3 Disturbed mice behavior rhythms before the Kobe Earthquake

Before the Kobe Earthquake Professors K. Nagai and K. Yagi were recording unusual animal behavior in knock-out mice at the Institute for Protein Research, Osaka University, 50 km away from the epicenter (Yokoi *et al.*, 2003). These mice lack a gene called Fyn (coding for tyrosine kinase), which regulates receptors in synapse membranes. The absence of this gene makes them unusually sensitive to stimuli—including electric shock. This was a serendipitous observation showing unusual mice behavior without any laboratory-applied external stimuli. The data showed clearly disturbed patterns of day and night activity before the Kobe Earthquake.

A mouse sleeps during the day and moves actively at night following the rhythm of an internal clock. Light resets the clock each day and the rhythm adjusts forward or backward as a result of exposures to light early or late in the night phase.

(a) Wheel-running activity

During the month of January, 1995, the wheel-running activities of four knock-out mice, one to a cage, were being recorded every 30 minutes. Their rhythmic activities followed a similar pattern up to until January 13 when they began to change, as in Figure 5.5 (a). The shaded background represents night (the active period), the

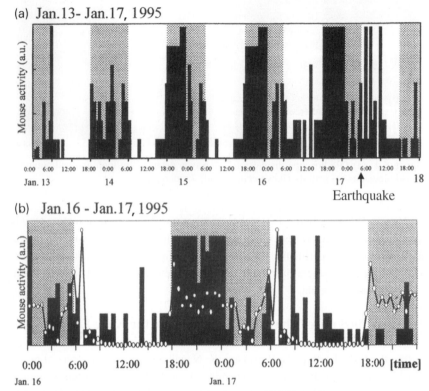

(a) Jan.13- Jan.17, 1995

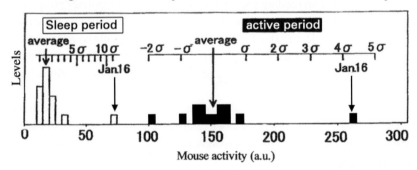

(b) Jan.16 - Jan.17, 1995

(c) Average levels of activity with standard deviations for January 16

Figure 5.5 (a) Increased levels of activity in mice after January 12, 1995, before the Kobe Earthquake at 5.47 a.m., on January 17. (Mouse activity shown in black; night (active period) shaded; day (inactive period) unshaded.) (b) Comparison of activities on January 16th and 17th with mean levels of activity (white circles) in January. (c) Average levels of activity in sleep and active periods and their standard deviations for January 16th. (Yokoi *et al.*, 2003).

dark blocks represent activity levels. An arrow indicates the time of the Kobe Earthquake. Figure 5.5 (b) shows a high level of activity on the nights of January 16 and 17 compared with levels averaged for night and day from January 1 to January 15, shown by the open circles. There were four small foreshocks on January 16 but they cannot explain increased activities on the night of January 15. Aftershocks might have affected the mice in the light period of January 17; the activity is higher than the average. (The mouse activity is compared in detail with observed ULF anomalies in Chapter 10, Figure 10.2.) One mouse was withdrawn from the experiment when its cage overturned and broke in the earthquake, but observations continued on the remaining three mice after the quake.

(b) Statistical presentation of mice activity

Figure 5.5 (c) shows average levels of activity in the mice for the period January 1 to January 15 and compares these with activity during the day and night of January 16. The standard deviations, σ, for both periods are also given. The activities on January 16 for sleep (day) and active (night) periods deviated by $+10.8\sigma$ and $+4.3\sigma$, respectively. The deviation for the totalled daily activity was 3.8σ. These are high deviations, as indicated in Table 5.2. The probability of a deviation of 6σ is one in two billion.

SD	$(m- SD)<f<(m+SD)$	$f > SD$
σ	0.68268948	0.31734
2σ	0.954499876	0.045500
3σ	0.997300066	0.0027
4σ	0.999936628	0.0000633
5σ	0.999999426	0.000000574
6σ	0.999999998	0.000000002

Table 5.2 The standard deviations (*SD*) and their probability shown by Gaussian distribution with the mean value *m*.

There are difficulties in dealing statistically with a rare event like a large earthquake, so the data for January 16 [Figure 5.5 (c)] were further analyzed with statistics of a single observation (Sokal and Rohlf, 1997). The normality test indicated that the probabilities of such unusual behavior were statistically less than 1 % (indices of 0.80 and 0.89) for sleep and active periods even using the small sample size sample limited to January. This kind of disturbance in activity has not been observed by the Institute for Protein Research in 15 years (almost 5500 days) of routine research activity using more than 10 mice—so that the probability that the behavior was chance is less than 1/55,000 = 0.0018 %. The mice behavior, certainly on January 16, was presumably in response to perception of certain preseismic signals.

Some seismologists were reluctant to accept the behavior of these knock-out mice as evidence of response to preseismic EM waves, and referees of papers submitted for publication sought similar evidence from other sites for statistical reasons. However these data were obtained serendipitously and the journals could not be persuaded to publish even on grounds of interest to other scientists who might have reviewed their old data in this new light. It took three years before the material found a journal prepared to publish.

5.4 An EM exposure experiment

To check whether EM waves were responsible for changes in behavior a mouse was exposed to EM pulses generated by electric discharges and its movements monitored by a video camera and recorded on a videotape. Two control experiments were set up: one mouse was housed in an acrylic box to protect it from charged aerosol in the atmosphere and another was electrically shielded from EM pulses by aluminum plates. Exposure to EM pulses during its sleep (light) period woke the test mouse up. Further exposure made it rub its cheeks and jaws incessantly—the same response mice made during the rock compression experiment—and burrow into the wood chips to hide itself. In the acrylic box the mouse behaved similarly. Isolated from the EM waves by the electric shield the mouse showed only slight initial responses while sleeping. So mice do respond to EM pulses in the absence of aerosols, and that is a sufficient explanation for their behavior.

5.5 Effects of EM pulses on chickens

Experiments: Electric discharges were made at a chicken farm using a Wimshurst generator. Small cages were placed in such a way that the generator was completely surrounded by chickens: 1000 within 50 m. At three meters the electric field component of EM waves was 8 V/m (magnetic field, 26 nT); at 50 m, 0.48 V/m (1.6 nT). When the pulses started, with discharge sounds like whipcracks, the birds suddenly stopped eating and chirping. They fell silent, poked their heads out of their cages, puffed up their feathers and watched the experimenter (Figure 5.6 (b). Then they began clucking noisily as after egg laying.

A watch dog at the farm ran from the site and rabbits also at the site jumped into the air as if shocked—consistent with reports of rabbit behavior before the Kobe Earthquake.

Reports of hens laying double-yolked eggs before earthquakes were not borne out in this experiment. The following day there were three eggs with a double yolk, nothing unusual. However, one egg was completely round in shape, something the farm manager had never observed before. It would be useful to monitor egg laying at egg-farms and correlate these records with earthquakes and thunderstorms for a possible picture of the effects of EM pulses.

The effect of generator noise on the behavior of the birds cannot be excluded, although EM waves of 1 W/m^2 at 250 MHz caused budgerigars to puff up their feathers and leave their perches (Section 5.2.2).

5.6 Possible physiological pathways for EM waves

5.6.1 Electrosensory organs in aquatic animals

Aquatic animals like sharks and catfish have extraordinarily sensitive electrosensory systems used for capture of hidden prey, for communication, for orientation and for navigation (Bullock, 1982). The electric field threshold capable of triggering a marine shark's feeding behavior is about 0.005 mV/m. The change in the electric field across the skin is transformed

Figure 5.6 Hens at a chicken farm (a) before electric discharges and (b) during discharges using a Wimshurst generator. After the pulses began all hens stopped eating and looked at the experimenter. Those closest to the generator appeared alarmed and retreated into their cages (TBS, TV program, *Amazing Animals* November 17, 2002).

into charges in the pattern of nerve impulses transmitted to the brain by electroreceptor cells similar to auditory hair cells. Figure 5.7 shows the electrosensory receptors as white dots on the skin of a catfish.

The electrosensory organs are divided into two categories: Lorenzini's Ampulla, a low frequency receptor for marine fish, and tuberous, high frequency receptors for fresh water fish (Figure 5.8, Bastian, 1994). Hundreds of individual receptor cells are grouped together at the base of an epidermal pit as indicated in (a). A canal filled with a jelly-like substance with a resistivity like that of seawater acts as a cable and helps communication from the receptor to the skin surface. The electric resistance along the canal, R_L, is much less than that across the wall, R_w. The wall capacitance, C_w, works as a shunt leading high frequency signals away from the

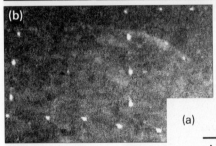

Figure 5.7 (a) Electrosensory receptors (the white dots) on the body of a catfish (b) Enlargement of (a).

receptor cells. These circuits might be simulated to make a catfish-equivalent sensor, (a).

The receptor cells lie embedded in the tissue below the ampullar floor. One of the receptor cells is enlarged in Figure 5.8 (b). The outside negative voltage causes a current to flow from the basal portion of the receptor and the voltage to drop. The inside of the cell becomes less negative, that is, depolarizes. This opens a voltage sensitive channel and allows Ca^{2+} ions to enter the cell. The calcium current further depolarizes the cell and increases the fre-

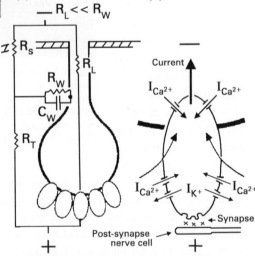

Figure 5.8 (a) An illustration of receptor cell currents for an electrosensory organ, *Lorenzini's ampulla* and an equivalent circuit diagram. (b) An enlarged receptor cell indicating both the calcium ion current, I_{Ca}, and potassium ion current, I_K. The x near the nerve cell represents neurotransmitters (*Physics Today* as reviewed by Bastian, 1994).

quency of impulses to the brain. A potassium channel is activated by intracellular Ca^{2+} ions. The calcium current, I_{Ca}, caused by the inflow of Ca^{2+} and counterbalanced potassium ions, K^+, flows outward after a short delay. This current causes electric impulses to the brain. The small Xs near the sensory neurons represent neurotransmitters released by the receptor cells.

If the outside potential is positive, rather than negative, the receptor's internal negativity increases, resulting in reduced frequency of nerve impulses. In this way fish detect small changes in electric fields.

The catfish has ampulla-type sensors to detect electric fields at a frequency range below 100 Hz. When one nerve cell connects with another, the transmission of an impulse takes place chemically by ions at the synapse. The speed is relatively slow because a chemical reaction of sodium ions flowing into the post synapse nerve cell is involved in the transmission and the time required for the chemical to polarize the post-synapse membrane and cause postsynaptic potential is about 0.1 ms. Hence catfish may not respond to an electric pulse with a width much less than 0.1 ms, as our experiments demonstrated. Animals behave in unusual ways when their nervous systems sense EM pulses with a wider pulsewidth.

5.6.2 Are EM pulses detected through the heart?

Most animals do not have evident electrosensory organs, so how might they detect EM waves? Possibly through their muscles, nerves and other organs acting as antennae. To test this hypothesis, a muscle—a live frog heart—was exposed to EM waves (Ohta, 2004).

Experiment:

A frog heart was extracted and maintained alive by circulating perfusion fluid through it. A single-turn loop antenna was set up and an alternating magnetic field applied at a frequency of 227 MHz (20 V/m) [Figure 5.9 (a)]. The beating frequency and the contraction force were determined respectively from the number of waves and their amplitude in the recorder chart, as shown in (b). These were constant before exposure to EM fields but decreased by about 10 percent after exposure. These results suggest that EM waves, presumably their electric field, polarized the membrane and inhibited the movement of ions. If that is the case, we may be able to say that animals detect EM waves through their hearts and speculate that signals from the heart will travel to the brain possibly causing secretion of neurotransmitters contributing to an excitement/anxiety response.

An immediate but transient increase in the mean arterial blood pressure and decrease in the heart rate were observed by exposure to an extraordinarily high power density of 33 - 65 MW/m^2 at 1.7 - 1.8 GHz (Jauchem and Frei, 1995). The researchers attributed the changes to the inherent hum in the transformer, but it is

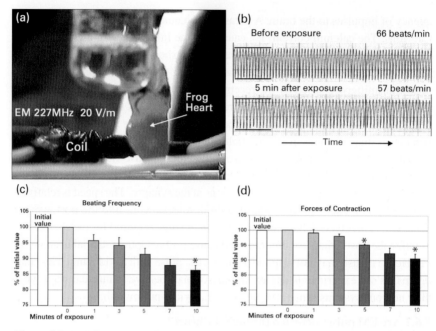

Figure 5.9 (a) A live, beating frog heart exposed an alternating magnetic field of 227 MHz (Electric field of 20 V/m, 1 W/m²). (b) Measured amplitudes and beat frequency before and after exposure. (c) Effect of EM waves on the heart's beating rate and (d) on the contracting force (Ohta, Ohani and Yamanaka, private communication.).

possible that EM waves may have been the cause. Our experiments, using waves of 1 W/m² (20 V/m) at 227 MHz did not generate any hum but did reduce the heart rate.

So, even muscles might be detectors—like Galvani's frog muscles.

5.6.3 Skeletal muscle with anisotropic resistivity

The resistivity of skeletal muscle perpendicular to an electric field is about seven times higher than that parallel to the field direction (Misakian *et al.*, 1993). Hence Japanese minnows, guppies, loaches and silkworms align themselves perpendicular to applied fields, to minimize electric shocks. The observed clustering of worms and snakes would be another behavior that reduces electric field effects (Ikeya *et al.*, 1996a).

5.7 Is behavioral response consistent with detection of EM waves?

How much reported unusual precursor animal behavior might be explicable as responses to EM waves? Probably much of that noted in reports retrospectively collected after earthquakes.

(a) Avoidance of water

The following behaviors are consistent with something passing through water that disturbs animals. Fish leaping out of aquaria and out of water on to land, ducks and crocodiles avoiding water, alligators retreating to forested areas away from rivers, sea lions at the Kobe-Oji Zoo leaving the water and behaving in distracted and unsettled ways on land before the Kobe Earthquake and crabs coming ashore.

(b) General response to irritation:

Mice, hamsters, cats, flies and ants—even turtles—experimentally exposed to EM waves begin rubbing, washing, preening, wiping behaviors, presumably in an attempt to relieve irritation. This is consistent with the restless, anxious, disoriented behavior reported of many animals before earthquakes. Reported early emergence from hibernation by reptiles and bees may also be associated with an irritant factor.

(c) Minimizing contact with the ground:

Cats have been reported to climb high in trees, chickens have flown to roof-tops, pigs have tried to climb walls, horses and cows have tried to stand on their hind legs before earthquakes. Thousands of earthworms appeared on the surface of the ground before the Taiwan-921 Earthquake. Swans lay on the snow with their legs in the air (indicating their legs were possibly their sensors). Apparently the uncomfortable entity was at its most intense at ground level.

(d) Unusual contact with metal:

Cockroaches were reported hiding under metal before the Izmit Earthquake, crows and sparrows landed and stayed on the hot metallic roofs of cars, or metal fixtures on buildings, a dog hid in an iron pipe with her pups before large aftershocks, flies stuck tenaciously to human skin. The animals may have been positioning themselves on conductive materials to minimize EM electric field effects. More probably they were seeking an electric shield.

(e) Body Alignment—usually in groups.

Flies rotating themselves on human skin, fish e.g. minnows, and silkworms, aligning themselves, dragonflies swarming in a single direction, swallows diving on a single axis to the walls of buildings, attempts by horses, pigs and cows to stand on their hind feet could be considered alignment behavior. All these behaviors might be an attempt by animals to minimize electric field effects by placing themselves at greatest muscular resistivity (perpendicular) to an electric field created by preseismic EM pulses.

(f) Animal behavior before thunderstorms: exposure to EM waves?

Some animals also show unusual behavior before thunderstorms, which also create EM fields. There is an old Japanese proverb about the weather,

It will rain when a cat washes her face

Lightning in thunderclouds approaching at a speed of 30 - 40 km/h would create an electric field of EM waves that cats could sense from a distance of 100 or 200 km, three to seven hours before it rains. Their sensors would be their eyes, whiskers, hair and their four paws in contact with the ground.

The pulse width in air of lightning from clouds to the ground is about one millionth of a second and the intensity of the electric field is very high: about 100 - 900 V/m. As mentioned, most animals are not sensitive to EM pulses of a width shorter than 0.1 ms unless the intensity is very high. Some lightning discharges among clouds and between clouds and the ionosphere have a longer duration (lengths close to 0.1 ms) and animals would detect the electric field and begin to rub their faces (Chapter 4).

Though anecdotal reports are controversial indicators of unusual animal behavior before earthquakes, we speculate that probably only EM pulses can travel through the ground, water and the atmosphere, and thus explain unusual behavior in aquatic, terrestrial and flying animals. We anticipate that bioelectromagnetic studies now being undertaken on the effect of mobile phones will provide further evidence of EM effects on animals, when animal size, frequency and direction of EM exposure are taken into account.

5.8 Waking before earthquakes: a human response to EM waves?

Many people said they woke ahead of the Kobe Earthquake. It is easy to conclude this could have been due to P-waves occurring before the main shock. For example, several pupils who lived at Kyoto, Okayama and Shiga, about 100 km away from the epicenter, were awoken by small vibrations and shouted, "Earthquake!", but there was no tremor. However about 10 seconds later, the main shock occurred. There is little doubt these individuals felt P-waves - which, in the epicenter area, should have arrived less than 10 seconds before the main shock.

However many seem to have sensed something else, but were reluctant to tell their stories, in case they were laughed at or teased for having extrasensory perception. Some said that they had woken up and were so scared that they got into bed with their mothers, or cried, or thought something awful was about to happen.

Because these stories were persistent and independent, a questionnaire was devised and handed to pupils in schools from Kobe through Kansai (Osaka and suburbs) to Nara and Nagoya in the east. The questionnaire asked students (8-10 years old at the time of the earthquake) if they had woken beforehand, and if so, how long before. With the exception of Kobe, in which pupils who had lost loved ones were not pressured to answer (about 5% refusals, and fewer in nearby Nishinomiya), 100% of pupils returned completed forms.

During interviewing of university freshmen in various centers, a similar questionnaire was completed, providing statistics for a wider range of areas, eventually from Kyushu in the south to Tokyo in the north. This had a 100% return rate, greatly reducing one potential source of survey error. More than 1000 questionnaires were returned from the two surveys. Although these students were relying on memories some years old, the incident was memorable and it is unlikely much invention occurred.

The results are shown in Table 2.1. They divide the data according to the wake-up time, particularly for the intervals < 1 min, 1-10 minutes, > 10 minutes. This is important, because if the children woke less than 1 minute before the main tremor, the cause could have been P-waves, but not if they woke up 1-10 minutes or > 10 minutes beforehand.

Table 5.3 The wake-up ratio (percentage) of pupils who lived at various distances from the epicenter and the time of waking (t *in* min) before the Kobe Earthquake.

City	Distance(km)	Woke-up/Total (%)		$t >= 10$	$10 > t >= 1$	$1 > t > 0$	$t = ?*$
Kobe	15	39/110	(30 %)	0	8	16	15
Nishinomiya	30	26/98	(27 %)	5	11	5	5
Suita-Takatsuki	40-60	25/83	(30 %)	6	11	6	2
Kitano, Osaka	42	36/182	(20 %)	5	22	3	6
Fukuzaki (West)	46	42/185	(23 %)	2	9	14	17
Nara (East)	75	36/187	(20 %)	3	22	10	0
Kyoto & Okayama	90-100	6/17	(35 %)	0	0	6	0
Nagoya	200	5/116	(4 %)	1	1	0	3
Hiroshima & Yamaguchi	400	2/55	(4 %)	0	0	2	0
Shikoku & Kyushu	400	0/32	(0 %)	0	0	0	0
Tokyo	500	0/31	(0 %)	0	0	0	0

* Wake-up times were not given

Perhaps the most striking initial result is that a surprisingly high number (20%) of children within about 100 km of the epicenter woke earlier than one minute before the quake, whereas for those more than 100 km away, the percentage waking early was close to zero. What could be waking these children? An appropriate chi-squared statistical test (Whitehead *et al.*, 2003), which combined zero wake-up reports with others, gave a very high value of 49, and calculation showed this result would only arise by chance about one time in a billion. This means these results are

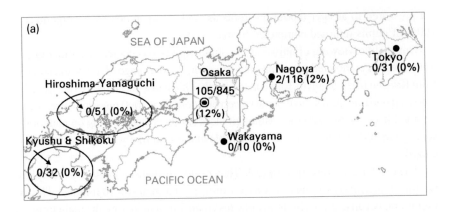

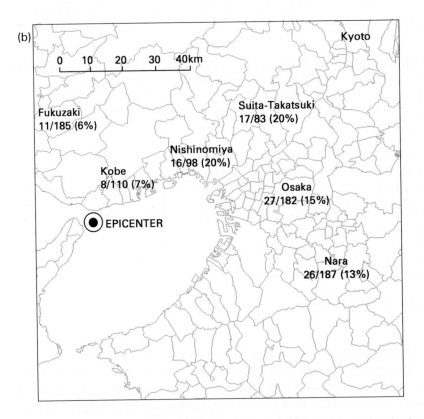

Figure 5.10 (a) A map of Western Japan and (b) the enlarged Kansai area showing proportions of student respondents who woke up more than one minute before the Kobe Earthquake. Kobe results are incomplete, see text.

not random and that there is a real effect differentiating reports closer to the quake from those further away.

Restricting the data to those children who woke earlier than 10 minutes before the tremor, a very similar analysis gave a chi-squared value of 22 and the chance of these being a random set of data was only 2 in ten thousand. This is still very significant for a chi-squared test, though not as dramatic as the first.

The results are displayed graphically in Figures 5.10 (a) and 5.10 (b) for close and distant locations.

If a psychological factor had woken the children, the percentages would have been uniform at any distance from the epicenter. However the chi-squared test shows a clear effect for the time interval 1-10 minutes before, and within several tens of kilometers from the fault—well before the arrival of P-waves.

This time interval fits well with some EM disturbance from near the epicenter, which travels much faster than other physical or chemical phenomena, and would be consistent with ULF or similar waves. This analysis cannot supply further detail of exactly what it was the children were responding to, except to suggest that it was anxiety-producing for children, as it seems to have been for many animals.

There was no foreshock on the morning of January 17. However, anomalous EM signals were reported to have been observed around 5 a.m. and just before the earthquake (Section 9.3, and Figure 11.2).

The electric field intensities estimated from reported unusual animal behavior at these areas ranged up to 70 V/m to 100 V/m as shown in Figure 4.22. One can speculate that this 20-30% group, aged eight to nine years at the time, was woken by EM pulses of about 100 V/m (25 W/m^2, 300 nT) at the epicenter area. The wake-up ratio in the west was small, in accord with the few reports on unusual animal behavior from those areas.

So, it appears that humans too may be detecting some precursors, possibly EM pulses.

5.9 Possible action of melatonin (the sleep hormone)?

Pineal melatonin rhythms are disturbed by an exposure to an intense electric field of 39 kV/m at 50 Hz but recover in less than three days after the exposure stops (Wilson *et al.*, 1986). Inhibition of melatonin production is known to lead to waking. One could speculate that a melatonin-inhibiting action is triggered by exposure to EM waves before earthquakes, so that wakefulness before earthquakes and increased activity levels in animals may be explicable as a melatonin reduction symptom. But such a link is highly speculative, given that EM exposure at 160 kV/m does not suppress melatonin production in rats (Tripp *et al.*, 2002*)*. A report from the Mayo Clinic says most insomniacs being treated with melatonin have not found it has helped much. Nor have any studies of melatonin effects thus far been able to rule out the effect of light. Clearly such a link can only be established by specialist investigation.

5.10   Connection between seismo-electromagnetic signals (SEMS) and animal behavior?

Pulsed EM waves have been independently recorded before earthquakes. Chinese seismologists observed pulsed electric fields before the Tangshan Earthquake of such intensity that their instruments went well off the scale. Unfortunately they discarded the data believing their electrometers must have been faulty to have produced such readouts (Private Communication, Wang, Anhui Seismological Bureau). Fraser-Smith (1990) observed EM waves at ultra low frequency (ULF) (0.01 Hz) before the 1989 Loma Prieta Earthquake in the United States (See Section 11.3.1 for more detail). Magnetic anomalies of 100 nT before a big earthquake (Rikitake, 1978) would correspond to an electric field of 30 V/m.

Chapter 10 looks more closely at the relationship between SEMS and animal behavior but our experiments indicate that unusual animal behavior starts on exposure to an electric pulse field of 0.2 V/m at a duration of 0.1 ms, and animals, such as mice and hamsters, start grooming at a few V/m (about 0.01 W/m²). An intense electric pulse of 2 V/m was detected simultaneously with observed sudden movements in eels and shrieking of hamsters before a nearby earthquake (Ikeya *et al.*, 1997).

Unusual animal behavior and other precursor behavior may be closely connected with the generation of EM pulses. More specifically, such animals may be detecting the pulsed current produced in stressed rock around a fault zone when dilatant areas are formed by microfractures ahead of an earthquake (See Chapter 3).

5.11 Summary

Rocks were compressed and fractured to simulate the microfracture (preliminary) stages of a small earthquake and to study animal behavior. Unusual animal behavior and EM pulses with an electric field of 2 V/m were observed during rock compression with the detection of acoustic waves ahead of granite fracture. Enhanced noradrenaline and reduced adrenaline and serotonin were observed after rock fracture in brain fluid and blood of rats placed beside the press. These changes were different from reaction to stresses caused by heat, sound and light, but similar to reactions to EM waves and to charged aerosols (though adrenaline increased in the last case). Though the interpretation is not clear there appears to be a stress reaction to EM waves in some animals that causes excitement. Mice at the Institute for Protein Research, serendipitously observed before the Kobe Earthquake, showed clear evidence of increased activity for several days before the quake and especially on the preceding day. A frog heart clearly responded to EM waves, and possibly 1000 chickens at a range of 0.48 V/m to 8 V/m. Twenty to 30% of children within about 100 km of the epicenter of the Kobe Earthquake were woken up by something (not seismic P-waves) more than a minute before the quake. The mechanisms by which animals respond to EM waves are not clear, and at this stage can only be speculated upon. But it does seem clear that they do respond to EM waves under experimental conditions and their behaviors are like those that have been reported to have occurred before large earthquakes.

6

Unusual Plant Responses
Before Earthquakes

Cut flowers in a container on the high voltage sphere of a Van de Graaff generator (and adjacent hanging ivy) respond to the movement of an electrically grounded rod. The movement of flower heads and leaves in still air—reported before the Kobe earthquake—can be explained by the appearance of electric charges on the ground. Video photo: the author demonstrating the EM effect at a public science show.

138

6.1 Introduction

Japanese legends suggest that plant growth and development might be affected before earthquakes: some plants are said to grow vigorously, others to bloom early, still others e.g. pear, apricot and peach trees to re-bloom before earthquakes.

Before the Kobe Earthquake flowers and leaves were observed to have swung subtly and trembled in still air. Before the Kanto Earthquake in 1921 the rice harvest in the area was reported to have been early and the plants stunted.

The mimosa is said to close its leaves before earthquakes.

Is it possible that both plants and animals respond to EM waves before earthquakes?

Plants have long been studied in connection with electricity. The direction of root growth is known to change near a large stone, because it disturbs the electric field in the soil. An interesting experiment was undertaken during the 1999 space shuttle mission to observe the growth of roots under an electric field without gravity. According to news reports, the root grew along the direction of the electric field.

In this chapter, some preliminary experiments are described to test the theory that electric fields of EM pulses, or EM waves at ULF—which we believe appear before large earthquakes—cause unusual plant effects.

6.2 Possible causes of plant anomalies before earthquakes

EM waves aside, what else might possibly produce the kinds of changes reported in plants before earthquakes?

6.2.1 Changes in the plant's environment

(a) Weather changes:

There is an old saying that weather before a large earthquake can be hot, rainy or very humid.

Humid and hot weather portends a huge earthquake.

This has not yet been established scientifically, but even if it were it is unlikely that several days of hot and humid weather before a large earthquake could cause the degree of change in plant physiology required for plants to grow vigorously or bloom early.

(b) Changes in the soil

(i) Soil porosity:

South American Indians believe that large earthquakes improve the fertility of the soil. Some Japanese earthquake prediction groups believe that the accumulation of seismic stress, microfractures and subsonic waves softens the soil and enhances root growth, resulting in good crop and fruit harvests. However, neither subsonic waves nor surface microfractures have been observed to soften soil.

(ii) Soil fertility:

Natural fertilizer produced by lightning might increase soil fertility. A Japanese proverb says,

> *Abundant lightning means an abundant rice harvest.*

Kenji Miyazawa, a chemist (also a poet and the author of the fantasy story *Milky Way Railroad*), wrote that lightning produces gaseous nitrogen oxides which become nitric acid and dissolve minerals to produce potassium nitrate and sodium nitrate, both fertilizers in paddy fields. Lightning is therefore called "Ina-zuma" in Japanese, meaning a wife (tsuma) of a rice plant (Ine).

Earthquake-associated lightning is coseismic and comes too late to have any effect but charges produced in soil by seismically induced EM electric fields could create small corona discharges at the tips of leaves and production of small quantities of paddy field fertilizer. (Small corona discharges occur at the tips of the tallest leaves of rice plants in paddy fields during thunderstorms, a miniature version of Saint Elmo's Fire—a faint glow caused by electrical discharge and reported around masts of ocean-going vessels or church spires during thunderstorms.)

(c) Removal of pests

Tributsch (1982) suggested that airborne ions and charged aerosols, which emanated from the fault zone, inhibited the proliferation of plant pests. This is possible in principle but might also eliminate beneficial bacteria and insects.

6.2.2 Changes in plant physiology: direct effects

(a) Beneficial effects of radon

Low-level radiation, slightly higher than the natural background, is said to stimulate biological structures and their physiology. Microcrack formation in underground rock before earthquakes allows the emanation of the radioactive gas radon, whose alpha activity affects living organisms. Roots and leaves close to the ground would be exposed to radiation from radon and its daughter nuclei and these plants might grow well due to this beneficial effect of radiation known as *hormesis*.

Atmospheric radon activity at the epicenter area increased considerably before the Kobe Earthquake but, again, for it to have the kinds of effects on plants described in reports it would have to take place well beforehand. In addition the nature of hormesis itself is still controversial; any effects would have to be distinguished from changes in soil porosity caused by ground fractures, and would probably be marginal.

(b) Hormonal changes stimulated by preseismic EM pulses?

When plants are exposed to EM pulses or charged aerosols, electric current flows in the conductive leaves, stems and roots, possibly stimulating their physiology (See following sections). This may contribute to early flowering and maturity. If the stimulation is too great, however, plants may wrinkle and die as described in legends and retrospective reports.

6.3 Investigation of electric field effects on plants and soils

6.3.1 Drying of soil through intense ground charges

Plants have been reported to wilt or die before earthquakes. *Coleus* (flame nettle) plants were reported wilting in Istanbul, 100 km away from the epicenter of the Izmit Earthquake, 15 hours before the quake, though the plants were healthy and the soil moist. Begonia plants, though growing new shoots, were also reported wilting through unusual loss of water (Ulusoy and Ikeya, 2003).

Changes in ground water level (produced by seismic stress) have been offered as an explanation for the withering and wilting of some plants before earthquakes, and this is no doubt true, but there is an old saying,

The soil in Bonsai pots becomes unusually dry before an earthquake.

The drying out of soil in ceramic pots has nothing to do with ground water.

Experiment on soil in pots:

Figure 6.1 shows the weights of two pieces of soil-filled pottery. A high voltage was applied to one pot using a Van de Graaff generator, while the other was kept as a control and no voltage was applied. The gradual decrease in weight shown by the upper black line is due to natural drying of the soil. The decrease is steeper in the pot to which the voltage was applied.

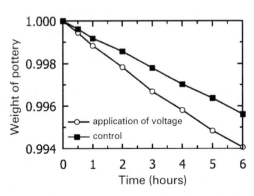

The rate of water evaporation is dependent on the applied voltage. Water molecules on the surface of soil minerals can easily be evaporated when the absorption energy of the soil is reduced by an intense electric

Figure 6.1 A graph showing the greater drying effects of voltages on soil in a plant pot compared with a control.

field or by the presence of charge. This is due to the electrostatic reduction of the *wet angle* between the mineral surface and the attached water (See Figure 8.6 for a similar effect in mercury).

So the bonsai proverb is consistent with the appearance of an intense charge before large earthquakes. Possibly, ultra low frequency EM waves induce an intense charge at the epicenter area leading to a high evaporation rate. Such a high level of evaporation could also account for the alleged high humidity rates before earthquakes as often described in the literature, and perhaps some anomalies in plants.

6.3.2 Visible and near-infrared images from a satellite

Preseismic activities alter characteristics of soils such as soil moisture (Sugizai *et al.*, 1980), gas content and composition (Rikitake 1976).

Images of thermal anomalies on the Earth's surface could prove to be a useful tool in scientific investigation of seismic processes by Satellite. Soil temperature anomalies of 2.5° C were measured before the Tangshan Earthquake (M > 7.0) in 1975. The NOAA satellite (North Oceanic and Atmospheric Administration, Tronin, 2002) observed anomalies of up to 3° C in the midst of stable background conditions 7-14 days before an earthquake. An earthquake in northern China in 2003 followed observation of an anomalous increase of the surface temperature in the desert in Kazakhstan, several hundred kilometers from the epicenter.

6.3.3 Unusual mimosa movements

(a) The closing of leaves and bowing of leaf stems

Mimosa pudica is called the *sensitive plant* in English and humble "bowing grass" in Japanese. A floral term for mimosa is "sensitive heart". The plant closes its leaves and droops if it is gently touched with a finger or when bees or insects approach. Japanese involved in earthquake prediction make the claim,

Mimosa close their leaves and bow before large earthquakes

Experiment: EM effects

A mimosa in a pot was placed on the high voltage sphere of a Van de Graaff generator as shown in Figure 6.2 (a). It closed its leaves when a high voltage (250 kV) was applied, as shown in (b) and the leaf stems drooped. When the sphere was discharged by touching with a grounded rod, the stems drooped further. When it was charged up again by removal of the rod, the leaf stems, with leaves still closed, pointed upwards. Electric discharges in the vicinity also made the leaves close and leaf stems droop. After about eight minutes the leaves opened and the plant returned to normal.

Figure 6.2 Mimosa on the sphere of a Van de Graaff high voltage generator (a) before and (b) after application of 250 kV. The plant closed its leaves on application of the voltage (Ikeya *et al.*, 1998). (Video photo.)

The author has grown mimosa for several years in pots in his room for use in demonstration experiments. The room is 30 m away from a bus terminal. When some bus engines start EM discharges cause flashing lines on the computer monitor. In some cases, the computer hangs. At the same time, on several occasions the mimosa leaves have closed and the stem has become limp. The sensitivity difference for different leaves may depend on the current induced in the leaves and the angle between the electric field direction and the leaf pulp.

Explanation: electrophysiology of mimosa

Electrostatically induced electric current flowed through the conductive stem after the appearance of charge on the Van de Graaff sphere. Even electric current from the air to the sphere through the stem made the mimosa close its leaves and bow.

The contraction movement of cells in *Mimosa* has been thoroughly investigated in electrophysiological studies (Toriyama, 1966; Toriyama and Jaffe, 1972) and the detailed mechanism by which the leaves close has recently been clarified at the molecular level. Just as the phosphorylation of the amino acid tyrosine in animal protein acts as an on-off switch in biochemical pathways, tyrosine phosphorylation of a contractile protein called actin controls cell shape in the mimosa. The *actin,* which forms a net and supports the cell from inside, loses phosphoric acid by electrical stimuli, which makes the leaves and stem shrink and bow. An alternative explanation (Kameyama *et al.*, 2000) is that the leaves close because of loss of water from the cell membrane.

(b) Old works on Mimosa by Bose

Dr H. K. Kundu, our colleague in the Gujurat work, noted work on the movement
of *Mimosa pudica* by J. C. Bose (1913) in Calcutta, India, early in the 20th century. Bose set out to record the reaction of mimosa to a number of stimuli—
including electrical stimuli. Three of them would have produced EM waves: electric shock from an induction coil, the making and breaking of a constant current,
and discharge from an electric condenser.

The apparatus in Figure 6.3 (a) shows how Bose measured the motion of the
plant. The recorder consisted of a balanced horizontal lever (V) at an axis (X),
supported on a frictionless jewelled bearing, and a vertical writer (W) whose end
was finely bent to press lightly against the smoked surface of a glass plate. A point
of the petiole of the responding leaf is attached to one end of the lever, the other end
bearing a small weight. The glass plate is allowed to fall at a uniform rate by means
of clockwork and trace a curve giving the relation of time to the mechanical
response.

Excitation at the leaf (A) was transmitted to the responding stem at B and the
drop in the stem pulled down the attached arm of the lever.

So the mimosa's response to EM pulses was established almost a century ago.

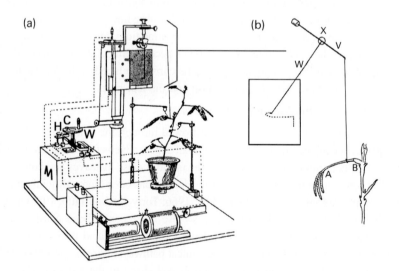

Figure 6.3 (a) Diagrammatic representation of Bose's apparatus to measure the
movements of *Mimosa*. M: spring motor, W: winding disc, C: projecting catch,
H: release handle for mechanical movement of the smoked glass plate (Bose,
1913). (b) A close up of the response recorder.

6.3.4   Other plant movement: "dancing" plants

A pot containing a lettuce plant was placed on the metal sphere of a Van de Graaff generator. When a high voltage was applied to the plant by removing the grounded rod from the sphere, the plant became erect and the leaves moved into a vertical position. When a leafy branch of azalea was placed on the sphere, the distance between the leaves increased and then returned to normal when the grounded rod was returned to the sphere.

Approaching the plant (or cut flowers) on the sphere with fingers or palm extended made the plants move toward them. A vase of cut flowers and trailing ivy hanging nearby also followed the movement of a grounded rod waved near them. (Students are delighted by these Tai Chi and "dancing plant" demonstrations, based on elementary physics.)

Explanation: leaves as conductive metal foil

A plant is an electric conductor and can be charged. The charge induced in the leaves and the opposite charges in the grounded rod (or the palm of a hand, which is essentially a grounded rod) attract each other leading to the movements of the leaves.

The plant responses, due to the electrophysiological response of the plant cell membranes and to electrostatic repulsion, were reproduced experimentally by applying a high voltage to the soil using a Van de Graaff generator.

We hypothesize that intense pulsed charges appear on the ground at epicenter areas before large earthquakes. In other words, ultra low frequency (ULF) EM waves (magnetic field about 330 nT; electric field 100 V/m) come out of the ground temporarily generating a flow of intense electric charges of the order of 1 nC/m^2. These could account for reports of unusual plant and leaf movements before earthquakes. For instance, subtle swinging of orchid flowers and fluttering of tree leaves in still conditions were observed before the Kobe Earthquake. Before the Taiwan-921 Earthquake, a co-worker's relative at the northern edge of the Chehlungpu Fault zone said she saw the unusual sight of plant stems and leaves in her garden pointing to the sky "as though they had raised their hands".

Intense EM waves would make leaves flutter through associated high voltages in the same way that electrode foils in a classical gold-leaf electroscope open and close as it is charged and discharged.

6.4   Silk trees in earthquake prediction

Botanically, a silk tree belongs to a species similar to Mimosa. Toriyama (1992), a specialist in the physiology of the sensitive plant, mimosa, learned that lay people

(a)

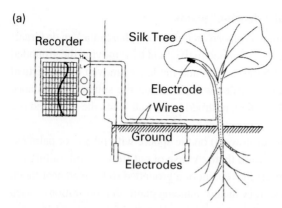

Figure 6.4 (a) Measurement of earth potential using the roots of a silk tree as a pathway for electric current. (b) Recorded "biopotential" data five days before, and the day of, the Miyagi off-shore earthquake (M7.4). Quake and aftershocks larger than M>3.1 are indicated by arrows (Toriyama, 1992).

(b)   Biopotential of a silk tree five days before a M7.4 earthquake

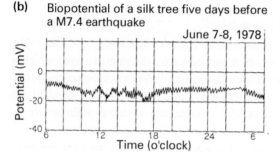

Biopotential of a silk tree on the day of an M7.4 earthquake with aftershocks of M>3.1

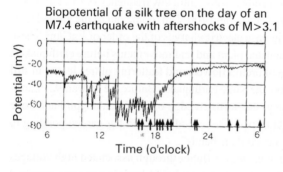

were using it in attempts at earthquake prediction. Having already reported the unusual response of mimosa before a typhoon, he decided to use silk trees to measure earth potential (differing earth voltages).

In his system the root of the silk tree acts as a pathway for the current, and a metal electrode is attached to the branch and another placed in the soil, as shown in Figure 6.4 (a). Toriyama detected pulse-like electrical noise and shifts in voltage levels over a long period of observation and argued for the use of such "biopotential" in earthquake prediction.

Toriyama successfully predicted the Miyagi Prefecture off-shore earthquake (M7.4) and for more than 20 years his "silk tree network" in the Tokyo area has had an 88% success rate in predicting earthquakes of magnitudes around M7 and epicenter distances of less than 400 km.

Toriyama's biopotential records before the Miyagi Earthquake are shown in Figure 6.4 (b) and (c), together with a diagrammatic representation of his silk tree system. In his Japanese book *Can Silktrees Predict Earthquakes?* he states: "Only weak signals were observed five days before the earthquake, but intense pulses were observed a few hours before the Miyagi prefecture off-shore earthquake (M7.4) with casualties of 28." The focal depth was 40 km, the epicenter 80 km off the coast and about 350 km from Tokyo. Arrows in the bottom chart indicate the main shock and aftershocks of M>3.1.

Professor H. Toriyama is presiding over a *Workshop on Earthquake Precursors* involving lay citizens and electrical engineers and may be measuring the earth potential (DC), EM pulses (and ULF waves) as silk tree "biopotential". Most seismologists are skeptical and the author has two concerns viz. comparison of data is difficult between trees because the roots (electrodes) are all different, and the electrode itself grows day by day so that comparison of data from the same tree is difficult.

Physicists are inclined to think that if a silk tree is really so good, an electrode should be made similar to the root of the silk tree, and similarly, that sensors like those found on catfish should be developed rather than using catfish themselves. Certainly stable electrodes used by geophysicists, especially by those who are using the VAN method (DC potential measurement), described in Chapter 11, seem better than silk trees. However, simultaneous measurement of earth potential by the VAN method at an experimental silk tree site indicated that the silk tree produced a smoother signal, according to Uyeda and Nagao who advocate the VAN method in Japan (private communication). Until geophysicists can come up with better detection methods, silk trees—like Custom's drug-detecting sniffer dogs—appear to have their place.

6.5 Effect of EM waves on the growth of rice plants

6.5.1 Unusually short rice plants before earthquakes

An early rice harvest of unusually short rice plants was reported to have occurred at the epicenter area of the Great Kanto Earthquake in 1921 (Kamei, 1978). The plants were clearly mature, but they were shorter than usual. To test the possible effect of preseismic EM waves in this anomaly, we exposed a species of rice, *Oryza sativa,* commonly grown in Japan, to EM pulses to see its effects.

Experiments on applied electric pulses:

(a) Experiment I:

Experiments were carried out in a biotron (an environmentally-controlled cabinet) fitted out with fluorescent lamps and air-conditioner. Temperature and humidity were set to 30° C and 80 %, respectively. Rice seeds were planted to three trays, two with aluminum electrodes and one without. Pulsed electric fields of 30 V/m were applied with two different pulsewidths, 0.5 ms and 5 ms, one to each of the two experimental trays, at a rate of 10 pulses/s for 12 hours during one period of light exposure, before germination. This did not raise the temperature of the tray. The results are tabulated in Figure 6.5 and Table 6.1.

Table 6.1 Effects of pulsed electric field on germination and growth of rice plants.

Experiment		Pulse width	Germination	Survival	Comments
I		Control	73-76 %	36 %	
		0.5 ms	80 %	10 %	Germinated plants wrinkled
		5 ms	46 %	10 %	Germination & survival plunge
II	Upper	Control	-	99 %	More light than lower control
	Middle	0.5 ms	-	72 %	Survival reduced/plants shorter
	Lower	Control	-	98 %	

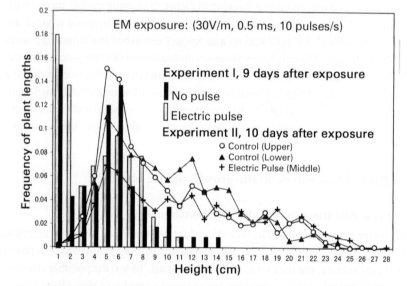

Figure 6.5 Rice plants exposed to EM electric fields of 30 V/m, showing effects on height. Frequencies are normalized to total number germinating for both Experiments I and II.

Pulses with width of 0.5 ms:

The rate of germination was slightly improved from 73 % to 80 % at a pulsewidth of 0.5 ms. Thirty six per cent of the control plants survived, but only 10% of the exposed plants. Nine days after exposure all the plants were removed, and for test and control trays separately, the numbers of plants at a given length were divided by the totals that germinated (to give the frequency of plant heights), as indicated by the bars in Figure 6.5. The length was generally short under the electric field, consistent with the report before the Great Kanto quake.

Pulses with width of 5 ms

Rates of germination were reduced from 76 % to 46 % and the survival rates from 36% % to 10 % after exposure to EM pulses with a width of 5 ms. Many plants died soon after germination. There are no known reports of poor germination at the earthquake epicenters, but this might be present if looked for.

(b) Experiment II:

One thousand seeds were planted to each of three trays placed at upper, middle and lower positions in the biotron. A pulsed field of EM waves at 0.5 ms was applied to the middle tray three days after germination (See Experiment II, Figure 6.5). The upper and lower trays were experimental controls and no field was applied. The lengths of the plants exposed to 0.5 ms pulses were measured 10 days after exposure to give frequencies of plants at a given length to germination totals (see connected data points in Figure 6.5).

The survival rate after germination was almost the same for the upper and lower trays (which had more and less light respectively than the centre tray), but reduced as shown in Table 6.1 for the exposed middle tray. Those that survived were wrinkled in the leaf. The shorter plants in the center tray were much smaller than in the control trays.

Summary

Clearly the lengths of the plants were affected by exposure to both pulsewidths— on average they were shorter than the controls. Although more plants germinated under 0.5 ms pulses, only 10% survived. At 5 ms pulsewidth germination rates almost halved. Generally the experiment had deleterious effects on the plants.

In Experiment I (Figure 6.5: exposure to pulses before germination) there was an excess of small plants and a deficit of large ones. In contrast, in Experiment II (Figure 6.6: exposure to pulses after germination) there was a deficit of small plants and an excess of large ones. This means that although exposure to electric pulses is generally negative for rice plants, there may be variable results depending on whether exposure is before or after germination—which is therefore consistent with reports of short but mature plants or of apparent vigorous growth, though the latter is at the expense of rather less yield, and may merely result from greater availability of light and nutrients for the surviving plants.

According to the literature, root development and early growth of radish were fostered by a pulsed magnetic field (Smith *et al.*, 1993). Because pulsed electric

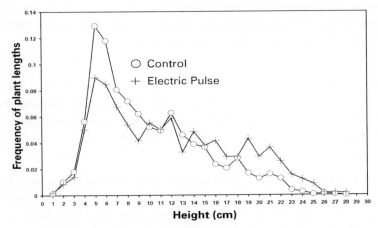

Figure 6.6 Experiment II. Rice plants exposed to EM electric fields of 30 V/m after germination. The two control points of Figure 6.5 have been averaged. Figures are adjusted so that the same number of plants are compared.

fields are generated by an intense pulsed magnetic field, we repeated the biotron experiment using radish seeds and found an increased rate of growth. In a later report on a replication of Smith's experiment, three species of plants were grown singly rather than multiply in pots (Davies, 1996), meaning there was neither competition between plants for resources nor interaction through their root systems. The new conclusion was that the increase in growth in radish exposed to an EM field in Davies' (and our) experiments appeared to be due to increased photosynthetic rate and accompanying influx of water. In one of our laboratory experiments, an exposure to EM waves also produced rapid growth of Shiitake mushrooms (*Letinus edodes*), concordant with a proverb,

There will be a good mushroom harvest after a year of lightning and typhoons

So, if EM waves emitted before earthquakes are responsible for short rice plants near epicenters and for more vigorous growth in some plants then these laboratory experiments lend support to that possibility.

6.5.2 "Bar code" in rice leaves

There is an old saying about an autumn weed called susuki, *Miscanthus sinensis* (Andersson) which has sharp leaves and belongs to the *Gramineae* family, as do rice plants. Yellow lines across susuki leaves are said to predict the numbers of typhoons. Similar yellow lines on rice plants are considered to be aids to earthquake prediction by some farmers who measure the positions of the bars and accordingly estimate the date of the earthquake expected that month. What might science have to say about the formation of such bars and any link to typhoons and earthquakes?

Experiment:

A high voltage electrode was placed horizontally above a tray of rice plants so that a vertical electric field was generated. Corona discharges were induced at the tip of the longest leaves. As the electric field intensity increased, the discharge at the tip resulted in browning of the leaves at the tip and growth

Figure 6.7 (a) The tallest leaves stopped growing after exposure to a vertical electric field between the soil and a ceiling plate. Younger plants then overtook them. (b) A photograph of the yellow horizontal lines generated by slight EM pulses (sub-lethal currents) in a leaf of an experimental rice plant (See also the front color plate).

stopped, resulting in wrinkling of the leaves. If the intensity was increased still further the leaf tip was burnt. Finally those leaves stopped growing and shorter ones caught up, eventually outgrowing the tall leaves with burned tips [Figure 6.7 (a)]. Slight discharges formed yellow bars on the leaves (Figure 6.7 (b) and front color plates).

Possible mechanisms:

(i) Chlorophyll:

Chlorophyll is one of the most important pigments involved in plant photosynthesis, the process by which light energy is converted into chemical energy for the plant's use. The chlorophyll molecule consists of a central magnesium atom surrounded by a nitrogen-containing structure called a porphyrin ring; attached to the ring is a long carbon-hydrogen side chain known as a phytol chain. Its structure is shown in Figure 6.8. Electrons of chlorophyll in green leaves are normally excited by light and are transferred from the magnesium base to the porphyrin ring and passed along an enzyme chain to produce a range of chemical compounds and sugars and an indirect greening effect.

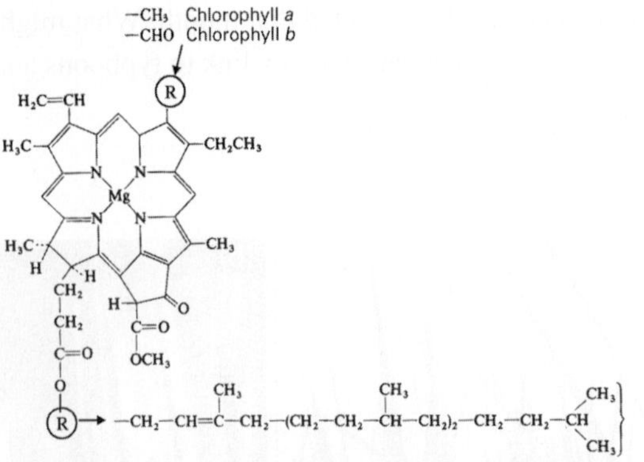

Figure 6.8 The structure of chlorophyll a and chlorophyll b with Mg at the center and a slightly different base, R. EM waves may interfere with the behavior of chlorophyll electrons during the process of photosynthesis causing yellowing and cell damage.

Electric fields could interfere with this process, throwing the electron out of the enzymic pathway prematurely (probably out of the longest point, the leaf tip), damaging cell structure and interfering with the laying down of green pigment. An electric field interference of several days may produce a slender yellow "bar code" on a leaf—the leaf resuming its normal color and health when the electric field effect stops. The more intense the electric field the greater the cell damage, creating burning of leaf tips.

(ii) Corona discharges:

Corona discharges generate nitrogen oxides and ozone. Leaves of sunflowers exposed to oxides of nitrogen or ozone show yellowing. Where the electric field is intense enough to generate corona discharges at the tips of taller leaves these gases would be produced, and may interfere with chlorophyll production in the photosynthetic process, laying down thin yellow bars across the leaf.

6.5.3 Discharges from the tip: Burning of plants at fault zone

After the Taiwan-921 Earthquake grass tips were found to be browned in a garden at the fault zone (Chapter 2). In a laboratory experiment when we applied a voltage of only 10 V to a mixture of water and soil in a plant container, the tips of the longest leaves became brown in color from discharges at the tip that injured the leaves.

The author ascribed the yellowing and wrinkling of bamboo leaves (the staple food of the giant panda) before the 1976 Tangshan Earthquake (M8.2) in China, to this electric field effect. (Giant pandas died from starvation as a result of effects on the bamboo.) Rikitake, in a friendly comment said that the drying and wrinkling of the bamboo were a result of changes in groundwater levels preceding the quake. However, the appearance of intense charges or EM pulses at an epicenter may also wrinkle the tallest tender tips of sharp leaves and produce yellowing.

Burning of plant leaves and roots at the fault zone were reported after the Tangshan Earthquake and attributed to frictional heat from fault movements. Alternatively they might indicate the burning effects of electric discharges induced by an intense ULF electric field of EM waves at the tips of leaves and roots as our experiments showed. The electric field in the ground needs to be only about 10 V to create small discharges and damage a leaf tip by burning.

6.6   Early flowering and re-blooming

6.6.1   Early flowering

For practical reasons we were unable to experimentally test the legends and reports of early blooming and re-blooming before earthquakes, but the obvious question to be asked is: do EM waves affect flowering cycles? A popular Japanese saying concerning plants in connection with earthquakes is,

> *If grass and trees re-bloom, a large earthquake will occur*

In the same way that EM pulses altered neurotransmitter levels in animals exposed to them (See Chapter 5), so hormonal pathways involved in flowering may also be affected by EM pulses. The flowering of the mustard weed, *Arabidopsis,* for example, is regulated by two floral inductive pathways: a day-light-dependent pathway which accelerates flowering on long days and a day-light-independent pathway which ensures flowering even if days are consistently overcast. It is believed that either activation of the floral inductive path or deactivation of a floral repressor leads to early flowering.

Further studies investigating possible activation by EM waves of the floral inductive path or an effect on the floral repressor would be useful—as would investigation of the effect of EM waves on florigen or ethylene. Florigen is a hormone-like substance that switches to flower formation from vegetative growth e.g. in the rice species *Oryza sativa.* Ethylene, in addition to inhibiting growth also promotes leaf fall and ripens fruit—in other words speeds up maturation, leading to early flowering. If such effects are observed it may be possible that preseismic EM pulses do have some effect on flowering and early crop maturity, as mentioned in reports and legends.

6.6.2   A new horticultural technology?

If EM waves or EM pulses are found to influence plant hormonal processes, they might be useful in cultivation, although it would have to be clear how much stress was being placed on plants by such exposure (Ikeya, 1998). EM pulses in such a scenario need only be small: widths of less than one thousandth of a second.

Researchers at Kinki University have succeeded in producing enhanced growth in mushrooms by exposing them to EM pulses generated by electric discharge.

However, plants with narrow leaves or tips might be harmed. Research would be needed to evaluate the risks and benefits of EM exposure compared with chemical fertilizers and pesticides, as well as nutritional changes.

6.7 Summary

Exposure of plants to electric fields of EM waves affects their growth and development. Electric fields of EM waves produced effects on plants consistent with those reported before earthquakes: short rice plants, yellow "bar code" effects on leaves, and leaf-closure and bowing of leaf stems in the mimosa plant. EM waves produced browning and wrinkling of leaf tips of rice, and at more intense fields, burning of tips of leaves and roots through electric discharges at the narrowest points. The mechanisms by which flowering, re-flowering and early maturity might occur in plants invite the efforts of biologists to establish possible EM effects on hormones, floral induction paths or floral repressors.

The silk tree shows a sensitivity to current created by differing charges at ground and branch levels—and, according to their reports, large earthquakes have been successfully predicted based on a study of its "biopotential", though scientists regard their results as random. Any results should be considered alongside those arrived at by SEMS detection methods discussed in Chapter 11.

If preseismic stress produces EM waves then plant anomalies noted in legends and retrospective reports may be earthquake-related.

Atmospheric Precursors

Earthquake Light, Clouds, Sun, Moon, Stars and Rainbows

Does fracture of the Earth's crust before fault movement affect the atmosphere? An intense electric field can explain legendary and contemporary reports of unusual atmospheric phenomena before large earthquakes. (Above): A tornado cloud generated at the tip of a high voltage needle electrode in a cloud chamber.

7.1 Introduction

Earthquake light (EQL) is various shapes and colors of light that appear on the ground or in the sky at the time of earthquakes, and sometimes before. The first earthquake light (EQL) ever recorded appears to have been in Rome in 373 BC (Musha, 1931); other historical accounts are mentioned in Chapter One. Tributsch (1983) also refers to historical reports and Papadopulous (1999) describes 30 cases of EQL reported in the East Mediterranean from the Fifth Century to the present. Musha also collected retrospective statements from witnesses of earthquake light at Izu, Japan, before the 1930 Kita-Izu quake (M7.3) and discussed their shapes. Yasui (1968, 1971) reported occurrences of EQL during the Matushiro earthquake swarm in 1967 and some are represented diagrammatically overleaf. Dai (1996) reported a locomotive driver who stopped his train when he saw EQL only moments before the huge Tangshan Earthquake (M 7.8) in 1976, which killed 240,000 people. His action apparently saved the lives of his passengers. Different types of EQL were observed before the Kobe (Tsukuda, 2000), Izmit (Ulusoy and Ikeya, 2003) and Taiwan earthquakes.

Not only light phenomena, but unusual cloud shapes have also been observed before earthquakes. Vapor trails (different from airplane contrails) are alleged to represent a type of earthquake cloud (EQC). Tornado (snake-like) clouds are another. See the front color plates for a tornado cloud.

Some lay citizens in Japan have designated themselves "Cloud Watchers" and attempt to forecast earthquakes by this means but meteorologists are skeptical that earthquakes in the lithosphere can affect the atmosphere in this way. Similar skepticism meets claims of other observed unusual phenomena before earthquakes: earthquake fogs (EQFs), red or haloed sun and moon, rainbows like short arcs and stars that appear closer and brighter.

In this chapter, EQL is explained as an electro-atmospheric phenomenon generated by intense electric fields at night, and EQCs and EQFs as vapor condensation similarly formed in a supercooled atmosphere. Earthquake phenomena involving sun, moon and stars may be caused by reflection and focusing of light when invisible vapor concentrations are formed by intense electric fields.

7.2 Earthquake light (EQL)

The morphology of earthquake light

Yasui, 1971, diagrammatically classified descriptions and photos of observed EQL (See Figure 7.1 and Figure 1.1.) His classes were: lightning with zigzag shape,

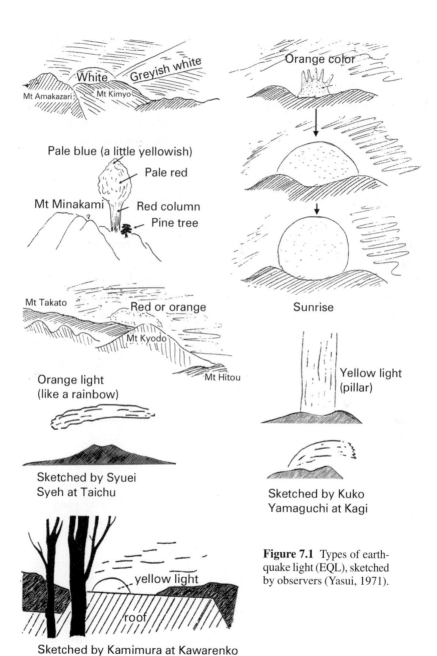

Figure 7.1 Types of earthquake light (EQL), sketched by observers (Yasui, 1971).

stroke lightning and sheet lightning, and various luminosities: swelling and shield-shaped, upward-extending funnel or trumpet-like forms, and belts of lights from an arc-like source. His zigzag lightning resembles ordinary thunderstorm lightning and also arc discharges from shorting of electric power lines, suggesting that EQL could be related to electric disturbance/activity in the atmosphere. Musha (1956), recorded early reports of fire balls appearing before earthquakes; more recently they were reported to have occurred before the Izmit and Taiwan-921 earthquakes in 1999. Ball lightning spheres with orange or purple-blue lightning haloes around them can pass through the wall or even through a body. There is a strong connection between ball lightning and weather phenomena such as storms and tornadoes, again possibly suggesting an electrical source.

7.2.1 EQL and the Kobe, Izmit and Shizuoka earthquakes

(a) Kobe

Flashes, domes of light and luminous funnels were observed just before the Kobe Earthquake. Forty kilometers from the epicenter, a graduate student and two faculty professors saw flashes of light a few seconds before the Kobe Earthquake.

Sources of luminescence and their development over time were studied based on information from witness interviews at 23 sites up to 50 km from the epicenter (Tsukuda, 1997, See Figure 7.2). They were estimated to be less than 200 m high. Horizontally they ranged in length from about 1 to 8 km. One light—an arc-like orange-colored light emanating from the eastern part of the aftershock area—was described as 1000 times brighter than moonlight. Some local flashing lights were

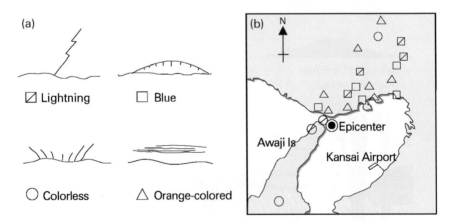

Figure 7.2 (a) Different colors and shapes of earthquake light (EQL) and (b) sites of EQL observed before the Kobe Earthquake (M 7.2) (Tsukuda, 1997).

reported. Luminosity started at ground level on land, suggesting that rocks near the surface were linked with the primary driving forces of the luminescence. Reported shapes included hemispherical dome-shaped light, sheet lightning and luminous cloud.

(b) Izmit and Shizuoka

People woken by the tremors in Izmit described white, blue, or sometimes green light at the time of the Izmit Earthquake. Forty kilometers away from the epicenter (at Adapazari) the light was observed as reddish or orange. At the time of the Japanese Shizuoka Earthquake, April, 2001, NHK TV broadcast a scene of the tremor from a camera on the roof of their building. (See Figure 7.3). Bright light appeared almost coseismically and decayed rapidly. Our analysis of the videotape indicated that almost all the light appeared in one frame only of the videotape so that the light lasted less than 1/30 second. There seemed to be two small

Figure 7.3 EQL captured, almost coseismically, in one frame (1/30 second) of videotape at Shizuoka, Japan (NHK TV program). (Video photo.)

sources of light in the immediately preceding frame. The local electric power company reported no occurrences of sparking power lines although there were electrically operated train services in the area.

7.3 Sources and causes of EQL

In this section we will look at possible mechanisms of EQL and discuss the likelihood of their involvement. However, we propose a "dark glow discharge" induced by an intense electric field (Ikeya and Takaki, 1996) as an alternative mechanism for EQL. The model will be discussed in detail in Section 7.4 but has yet to explain how such an intense charge can be generated and maintained in a conductive earth.

Three states of material are suggested to emit light: solids, aerosol, and gases or atmosphere.

7.3.1 Rock materials:

Solid state luminescence

(a) Triboluminescence

Luminescence caused by friction and observed during fracture of materials is called triboluminescence. Triboluminescence was suggested as a possible explanation of luminescence seen in landslides caused by rock fracture, and also of EQL (Terada, 1932). He subjected rocks to grinding by a revolving disc driven by an electric motor. The light emitted from the contact of rock and grinder was compared (in the early 1930s when no good standard light sources were commonly available) to that from 100 glowing incense sticks used on Buddhist altars.

The difficulty with solid state luminescence is that light effects need to be explained that are occurring well above ground level. Luminescence generated in rock tends to hug ground surfaces. As Terada (1932) comments: known mechanisms of solid state luminescence have difficulty in explaining EQL observed above mountains covered with trees, because even if air molecules are excited, diffusion rates are rather slow and the luminescence decays too fast.

(b) Thermoluminescence (TL) of debris

Figure 7.4 shows an illustration of the TL process. A pair of electrons in a solid mineral, represented by a boy and a girl, is separated by natural radiation: alpha, beta and gamma rays. One of the pair of charged electrons, represented by the girl, is trapped by an impurity, illustrated by a "gangster". Another unpaired electron (a depressed boy) is in a state of electron deficiency—called a "hole" in solid state physics. The trapped electrons and holes accumulate in natural minerals and their

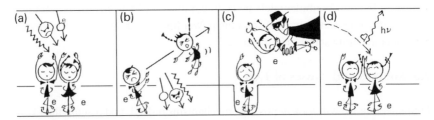

Figure 7.4 (a) A pair of electrons in a solid mineral. (b) Natural radiation ionizes paired electrons, represented by a boy and a girl, the girl being trapped (c) by an impurity (gangster). (d) The trapped electron (girl) is thermally released by heating of a mineral and recombines with a hole (boy). The reunion of boy and girl produces the light called thermoluminescence (TL) (Ikeya, 1993).

concentration is proportional to the age of the rock and the dose rate of natural radiation. Heat releases electrons from the traps and results in the recombination of electrons and holes, creating thermoluminescence (illustrated by the pleasure of the characters in the cartoon).

The movements of dislocations in minerals (Figure 3.15) during microfractures and fractures leading to an earthquake, might release electrons/holes trapped in minerals leading to recombination with holes/electrons and emission of light— hence earthquake light. The mechanism of this release of electrons might be frictional heating[1]. The question must be asked: could thermoluminescing dust spread into the air from the epicenter by underground rock fracture account for EQL? The temperature of the dust in this scenario would be elevated by friction so as to give TL. Solid carbonate materials should thus luminesce orange-yellow due to the excitation of Mn^{2+} involved in such carbonates. Quartz and feldspar would show blue and red luminescence.

However, for debris composed of grains of quartz and feldspar generated by granite fracture, temperatures would have to be higher than $250°$ C and $320°$ C, respectively. Not only that but the TL would have to last for more than several seconds, meaning hot dust would have to stay hot for many seconds.

But there is no evidence of high temperature debris emission from epicenters before or at the time of earthquakes. EQL also occurs in places where luminescent materials do not exist. However, TL of rock materials may account in some cases for a brief low-level glow.

(c) Luminescence caused by exoelectron emissions

As explained, stress and microfractures release electrons/holes that have been generated by natural radiation and trapped in minerals for long periods geologically. Some electrons are emitted from minerals to the atmosphere, and this is called "exoelectron emission". The electrons charge or excite the atmospheric molecules (Enomoto and Hashimoto, 1990). Whether this is identical to triboluminescence is not clear at present. As in the case of TL induced by frictional heat, trapped electrons, which are not stable at higher temperatures in the crust, must be present.

But in exoelectron emission the emitted electrons would have low energy and activity and any glow would be quite localized.

---

[1] Dating the event of the movement of a geological fault (an earthquake) using TL and Electron Spin Resonance (ESR) is based on this mechanism (Ikeya *et al.*, 1982; Ikeya, 1993). The clock time is reset to zero by faulting or fractures, and natural radiation thereafter generates trapped electrons again. The date of an earthquake can therefore be determined by measuring the concentration of trapped electrons accumulated after the clock is reset to zero by the movement of the fault.

7.3.2 Aerosol, liquid and gaseous processes for EQL

(a) Neutralization of charged aerosols

Tributsch (1982) attempted to explain all precursor phenomena with his hypothesis of charged aerosol emissions from the epicenter area, as discussed in previous chapters. He argued that EQL is luminescence from recombination of charged aerosol with electrons i.e. neutralization of the charged aerosol generates light. In this respect his hypothesis is closer to solid state luminescence of debris than a gaseous process.

Neutralization of ions and molecules and dust particles leads to excitation of atoms and molecules, which can be a potential source of light. However, a charged aerosol, composed of submicron to micron-sized solid particles floating in air like smoke, cannot emit visible light luminescence by neutralization. The electric recombination energy will be distributed to the large number of atoms constituting the aerosol, but as kinetic energy to jiggle each atom. There will not be enough energy to both excite the atom and to provide photon energy of 2 to 3 electron volts (eV) the minimum required to produce visible light. Hence, the hypothesis of EQL from charged aerosols is not tenable.

(b) Sonoluminescence

Luminescence generated by supersonic sounds from molecular reactions in water shaken by compressional (P) waves is called sonoluminescence. It has been suggested as the cause of EQL (Johnston, 1991). It may be a candidate for light known to be associated with tsunamis (blue and yellow balls of light are known to occur at the crests of tsunami waves), but can hardly explain EQL over mountains.

(c) Combustion of natural gas and decomposition of methane hydrate

Burning of natural fuel gases such as methane has also been considered as a cause of earthquake light or fireball lightning (Barry, 1981). Earthquake light above the sea, and also tsunami light, might come from gaseous methane released from gas hydrates in sediments under the deep ocean (Enomoto, 1999). Tsunami light and the reported blue-red light on the sea floor during seawater withdrawal after earthquakes might be explained by combustion and electric discharges in an atmosphere containing dilute gases.

(d) Recombination of electrolytic gases

We might speculate that in the case of tsunami light, the gas is hydrogen generated by electrolysis of water. Fractures caused by the earthquake would produce unpaired electrons on the surface of minerals, which would react with ambient water to generate radicals of OH and H. These would react further with water to generate gaseous hydrogen. The presence of hydrogen at a fault zone is known, as mentioned in Chapter 3, Section 3.2.5. Electric current flow will also decompose water, so both processes will generate gaseous hydrogen and oxygen. Explosive recombination of these would emit light which might occur even under the sea, like the bright balloons of light seen emerging from the Gulf of Izmit and the Eastern Marmara Sea before the Izmit Earthquake (Chapter 2).

7.3.3 Atmospheric light and mechanisms for EQL

(a) Radiowave reinforcement in ball lightning

There are numerous proposals on sources of energy for ball lightning, which is a major member of the class of EQL in the atmosphere. Skeptics argue that EQL is caused by sparking from short-circuiting electrical power transmission lines shaking in quakes. Generally, however, power lines are protected by insulators. EQL reports are also centuries older than power lines, and many EQL reports are preseismic. However, sparking of power lines during quakes undoubtedly occurs.

The Russian Nobel Prize winner Pyotr Kapitza (1955), famous for low temperature physics studies, claimed ball lightning was caused by electromagnetic radiation. However the energy required to maintain its continuous luminosity was greater than the fireball itself could provide. Because the dimensions of ball lightning (about 20-40 cm in diameter) are close to the wavelength of microwaves (30 cm for 1 GHz), Kapitza speculated that the outside energy might come from microwaves, and that ball lightning would most probably be found in regions where microwave intensity was highest. But a field expedition to an altitude of 2000 m in the southern Caucasus, where thunderstorms frequently occur, found only weak microwave production at a site far from the thundercloud—so the outcome was inconclusive.

Obviously the mechanism of interference and enhancement by natural microwaves produced in thunderstorms invites more research. Seismic EM signals (SEMS) may also be a source of such natural microwaves, and possibly, thus, a generator of earthquake related ball lightning. Kamogawa *et al.,* (1999) suggested an energy transfer from EM waves to microwaves of particular frequencies, a mechanism known as Anderson localization of the electron wave function in physics. Thus EM waves could explain ball lightning but there are many conceivable models for physicists to play with.

(b) Combustion of fuel gases and electrolytic hydrogen?

Natural gases may be involved in fireball lightning as they may be in tsunami light. The sound of an explosion reported just before the 1999 Izmit Earthquake might be explained by the explosion of fuel gases, but rock fracture can also generate sounds like explosions. Natural gas combustion and its discharges might produce ball lightning, but cannot not explain all EQL. One difficulty is that fuel gases cannot combust under water and at low temperature. Hydrogen might be generated by electrolysis at the surface of rocks or minerals from preseismic waves at ULF. Ignition of the hydrogen by electric charge could create a glow.

7.4 Theory of a dark discharge model

St. Elmo's Fire is a glow that appears at the tip of masts of ships or church towers during stormy weather, caused by brush-like discharges of atmospheric electricity created by high voltages. We propose here that such high voltage discharges are the origin of EQL. Scientists use the term "corona discharge" for this, and the principle is used for charging the drum of a copying machine or a laser printer. When the ions (charged atoms and molecules) produced are neutralized, the neutral atoms or molecules remain energetically excited and can give out this energy in the form of light. Both excited nitrogen and oxygen molecules in the air can and do emit such light, mostly blue, green and red luminescence. For further discussion see Section 7.4.2 and Table 7.1.

Such atmospheric charges may occur as EQL when an intense electric field is produced by the fracture of a massive block of bedrock granite hitherto blocking the movement of an active fault.

7.4.1 An electromagnetic (EM) model of a fault

In Geophysics, particularly Seismology, it is possible to construct a mathematical model of a fault. From its length and the amount it shifts in an earthquake one can calculate e.g. the earthquake energy and the magnitude of the earthquake. This depends on the rock properties, because a soft weak rock like mudstone may fracture with a very low stress, and hence frequently, whereas a much stronger rock like granite may accumulate stress and fracture rarely, and only when the stress becomes extremely high. The result of the latter is an earthquake in which a lot of energy is released. It is possible to calculate the relationship between the earthquake magnitude and other earthquake-related parameters for different rocks, and to derive the scaling relation (the rate the magnitude changes) theoretically (Ikeya *et al.*, 1997, 2001).

Such a theoretical derivation was used to explain EQL as an EM phenomenon by analyzing the piezoelectric properties of quartz, a common rock mineral. In its very simplest terms this is like the sudden stress on a quartz crystal in a gas lighter, which creates a high voltage spark. However the actual situation is much more complex and has unexpected features, particularly that high voltages can be generated at all in a rock, and second, that they are much, much more intense when the fracture is slow and local (as for preseismic activities) rather than in the main shock itself.

The detailed quantitative theory is given in Ikeya *et al.,* (1997) but this section attempts to explain some of the qualitative aspects and the consequences. (The quantitative theory shows that the charges or voltage produced by piezoelectric properties are higher by orders of magnitude than those generated by other mechanisms but lower than expected from the phenomenon in nature; some other unknown mechanism may be operative deep underground.)

The creation of charges by the piezoelectric mechanism in a gas lighter relies on a single crystal, which must be hit at the correct angle. In a rock, the difficulty is that, at first sight, there is no particular angle for the quartz grains, so that all charges, including piezo-compensating charges, are produced in random directions, and in such circumstances they all tend to cancel out, giving near-zero voltages. However the theory shows that even with completely random orientation in the quartz grains the EM wave intensity is not zero, but the charge pairs on both sides of the quartz grains, temporarily, at stress release, give a voltage proportional to the square root of the number of quartz grains (See Figure 7.5). In fact, even if only 1% of quartz grains are initially aligned, the intensity of the voltage created by alignment of piezo-compensating charges is considerable. If this is the case, the movement of charge in the rock will generate intense EM waves. As described in Chapter 5, experimental work showed that detectable EM waves were generated when blocks of rocks were fractured under

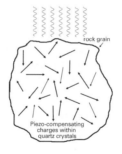

Figure 7.5 (a) Emissions from apparently randomly orientated piezo-compensating charges which help to create an intense electric field. Although the total dipole moment is zero, they emit waves as if only one big dipole were present.

high pressure, even though the stresses were extremely low and would not normally give a detectable earthquake. This is direct experimental support that the theory is correct and that stress will generate high voltages in nature.

It is important to note that the EM model applies both at the macro and microlevels; it is an appropriate mechanism for the production of EM waves both at the microfracture stage and at the time of fault movement (i.e. the earthquake).

The theoretical model was used to calculate the pulsed charges and emission of EM waves from a typical fault zone by preseismic stress. The model was homogeneous and the stress was released over a period of one second—obtained by dividing the half-length of the fault (5 km) by the speed of S-waves (4 km/s), which was considered appropriate.

The electric field was generated by pulsed charges in granite, and the results are shown in Figure 7.6, noting that the dielectric coefficient is ε and the resistivity ρ, with values from Takaki and Ikeya (1998). A remark should be made here about the absolute intensity of the field; it is orders of magnitude smaller for the main fracture in (a) than for the preseismic local fracture in (b), lending support to the likelihood that the preponderance of reported phenomena with EM origins will tend to be observed before earthquakes rather than coseismically.

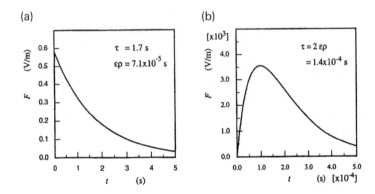

Figure 7.6 The time dependence of electric field intensity on the ground for the fracture of a quartz grain with a size such that the effective stress drop with time constant, τ, is about 10^8 N/m² (about ten thousand tons/m²) for a rock with resistivity ρ. (a) Coseismic faulting. (b) Preseismic fracture.

It is also worth noting that voltages as high as 4000 V/m can be generated by the model, which are enough to produce significant effects by various means at significant distances from the epicenter giving some perspective on the fact that animals often reacted to only 1 V/m in tests described earlier. Higher voltages are certainly possible.

The next section describes how these high voltages and the production of EM waves can promote production of light, EQL.

7.4.2 A dark discharge model for EQL

(a) High voltage formation

The creation of such intense electric fields on the ground may induce lightning strokes from clouds to the ground or between clouds—a different kind of EQL from that about to be discussed, but one of the class.

This part of the theoretical model relies on acceleration of electrons by high voltages. These electrons then strike gas atoms in the air, making them glow. This model has also been published (Ikeya and Takaki, 1996). There are normally 4-10 million free electrons generated in a cubic meter of air by cosmic rays and natural radiation each second and these can be accelerated.

How can voltages produce light? An intense electric field in a small space does not usually produce a glow in the air. For example there is no glow around a power socket in a darkened room even if the power is on. However scientific instruments can detect a feeble light at a much higher voltage. This light, which we shall call light from a "dark discharge", would be visible in conditions to be explained shortly.

In Ikeya and Takaki (1996) it is calculated that if sufficiently high voltages are present, electrons may be accelerated to very high speeds (94% of final velocity) within 10 nsec (ns: billionth of a second), and cause light emission even to heights above trees on mountains. This will work optimally if voltages are 60-150 kV/m. Alternatively a pulsed electric field of seismic EM waves might accelerate the electrons. In the model, charges induced by EM waves at ULF were approximated for charge densities of 120 microCoulombs/m^2 in the area of a local fracture 10m x 10m for 2 ms. This electric field is so intense that a kind of electron avalanche or chain reaction results in electrons hitting atmospheric molecules and producing ultimately more electrons and light. Some of the light is UV, which further ionizes atmospheric molecules. If it were possible to observe this air in a glass jar, the glow would not be visible, but if an observer at a distance could look horizontally through thousands of cubic meters of air above the fault, calculation shows that it would be visible as a brief glow from 1 km away, and sufficiently intense to be visible under a full moon. In laboratory tests of atoms excited by accelerated electrons, the light from the excited nitrogen atoms decayed with a time constant of about eight seconds. The luminescence of green oxygen lines (557.7 nm) increased but those of red oxygen lines (630 nm) and the OH-band did not change under electron excitation (Fishkova *et al.*, 1985). (See Table 7.1 for details of the colors and their origins.) This means that the quality of light would correspond to that seen near epicenters, but does not yet explain the red colors observers claim to have seen.

If a rock is fractured in the laboratory under mild stress conditions the glow is visible along the crack (and depends on the gas atmosphere composition). It is easy to imagine glows a million times more intense with the catastrophic conditions of a rock fracture in nature and hence this model seems convincing, though all the details of the intense electric field generation by underground rock fractures have not yet been established.

Further calculation established the shape of the glow as a dome or ball lightning shape, using typical known initial electron energy conditions (Figure 7.7). Appro-

Table 7.1 The wavelengths of emissions from excited states of N_2 and O_2 molecules as well as oxygen atoms, for earthquake light (EQL).

Source		Color of light	Wavelength	Transition levels
Nitrogen molecule	N_2	Blue and UV	361 nm	Level B_{3g} to A_{3u+} in 8 s
		Red	650 nm	Metastable A_{3u+}
Oxygen molecule	O_2	Blue light	400 nm	
Oxygen atom	O	Green light	557.7 nm	Auroral light
		Red light	630 nm	Auroral light
Argon (?)*	Ar	Red light		From fractured rock ?

*Rocks contain argon from disintegration of radioactive potassium-40. However, atmospheric argon is far more abundant than that released from fracture of rocks.

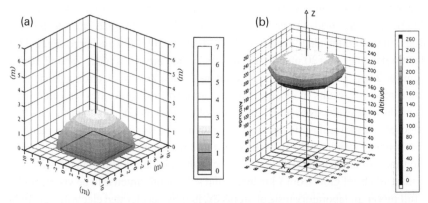

Figure 7.7 (a) The density of electrons in the center of the area where the charge appears. The spatial distribution of EQL is calculated by using a pulse charge density of 10^{-5} C/m² lasting for 2 ms in an area of 10 m x 10 m (Takaki and Ikeya, 1998). (b) UFO type EQL when the mean free path of electrons (average distance between collisions) is two orders of magnitude lower.

priately, since it looks like a UFO, it may move back and forth along the fault line as fast as 3-4 km/sec as the location of the stress release changes.

(b) Red color at a distance

The EQL (as reported before the Kobe and Izmit earthquakes) was white or green at the epicenter, and reddish further away from the epicenter. Where could red colors originate? Although red colors are seen in auroras at high altitude due to excitation of oxygen atoms, that mechanism does not work at the surface of the earth in much higher gas pressure, because other atoms frequently strike the excited atom and steal the extra energy. Another mechanism which must be ruled out is excitation of the 1% argon in the atmosphere. In theory extra radioactive radon gas emitted pre- and coseismically could excite red colors in argon by the emission of alpha particles. Although this would give a red color it would also do so at the epicenter, and it would take many hours for the radon to diffuse away, so this is not a likely explanation for a short-term EQL.

The explanation may be that red light travels much further in the atmosphere than the blue or green light, so that an observer a long way from the epicenter may not see the original green or white light but only remnant red. So reports of red light at the time of earthquakes are not necessarily superstitious or illusory, as some critics contend.

7.4.3 Lessons from ball lightning

Ohtsuki *et al.*, (2002) reported coseismic ball lightning moving along the fault line in the Kobe Earthquake and colliding with Mt. Rokko, followed by cloud-to-ground lightning. Many observers saw ball lightning on Mt. Jiufen before the Taiwan-921 Earthquake, which destroyed sections of the mountain. They resembled small volcanic eruptions: one on the top of the mountain and three near the top, leaving holes about 50 cm in diameter.

Many eyewitnesses have reported fireballs traveling along electric wires. They also move along an intense electric field even if there is an opposing flow of air. As already mentioned it has been suggested that the most favorable places for production of ball lightning may be regions where radio waves, first generated by thunderstorms, reach their highest intensity through interference with each other. Although electrical fields of more than 100 kV/m exist frequently in thunderstorms, there are not many quantitative observations of preseismic intense electric fields at an epicenter before a large earthquake. So, observation of electric fields before earthquakes is necessary if EQL is to be validated as an electro-atmospheric phenomenon. Chapter 9 offers indirect evidence of this.

Ohtsuki and Ofuruton (1991) used a microwave cavity as shown in Figure 7.8 (a) and produced small glowing balls a few millimeters across as shown in (c). The metallic looking spheres, surrounded by orange or purple-blue haloes, blew out of a cavity made of aluminum punched plate, as shown in (b). They also moved through a 3 mm thick ceramic board without causing any damage and left small circular imprints on a metal plate. A different type of plasma fire, as shown in (d), was observed 20 s after the first appearance of the fireballs.

The evolution in the scientific understanding of the nature of the fireball is an object lesson for those seeking to understand and explain not only earthquake light but also other earthquake related phenomena. After more than 400 research papers

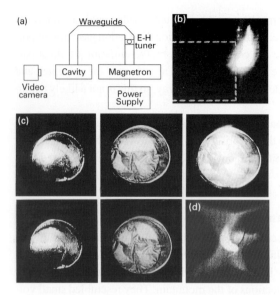

Figure 7.8 Successful reproduction of fireballs in a laboratory. (a) A diagram of the apparatus, (b) plasma which blew out of the cavity (whose outside limits are indicated by dashed lines), (c) plasma fire balls observed in the cavity and (d) the shape of the fireball 20 seconds after it appeared (Ohtsuki and Ofuruton, 1991). The shape (d) apparently resembles some observed EQCs.

and five books on ball lightning between 1970 and 2000, H. Tutt (1997) had this to say in his book, *Unexplained Natural Phenomena,*

The story of ball lightning has shown how, in some 30 *years, a 'fringe' subject, worthy of ridicule in the orthodox scientific community, can become a respectable, mainstream area of study.*

The scientific community needed to be convinced that fireballs were not merely the creations of superstitious and impressionable people. It is the author's hope that unusual phenomena observed before large earthquakes will, like the fireball, prove to be rare natural phenomena with a scientific basis.

7.5 Earthquake Fogs and Clouds (EQFs and EQCs)

7.5.1 What are EQFs and EQCs?

Figure 7.9 shows a cloud that suddenly appeared in a blue sky only moments before the Great Kanto Earthquake of September 1, 1923, that destroyed Tokyo. Official data put the time of the earthquake at 11.58.44 sec; the photo was taken barely a minute before. The cloud had a luminous white exterior. A similar cloud generated just before a large aftershock of the Kobe Earthquake was also photographed (Wadatsumi, 1995). Fogs were observed in the region around Nishinomiya City, east of Kobe, preceding the quake in 1995 (in fine weather) and seen covering a graveyard at Izmit at the time of the 1999 Izmit Earthquake.

Not surprisingly earthquake clouds and fogs have a significant legendary history. A Fifth Century Indian religious treatise Brihat Samhita discussed clouds and animal behavior as signs of earthquakes. Ancient Chinese and Italians tried to pre-

Figure 7.9 A photograph of an EQC that suddenly appeared just before the Great Kanto Earthquake on September 1, 1923 (Photographed by E. Ohshima; Kamei, 1976).

dict earthquakes on the basis of peculiarly shaped clouds. In the modern era the Mayor of Nara City sought to give credibility to old proverbs about "earth air" and clouds (Kagita, 1982), and formed the aforementioned group of cloud watchers now widespread in Japan. Tributsch (1982) devoted a chapter to EQFs, referring to other mentions by the Greek philosopher, Aristotle, and by Alexander von Humboldt in his *Journey into the Equinoctial Regions of the New Continent.* EQCs have been classified into different types in books (Kamei, 1976; Lu, 1981) although scientific communities have not accepted the existence of EQCs.

7.5.2 Shapes of EQCs

No statistical analysis has been done of earthquake predictions by cloud watchers to establish whether their results are random or statistically meaningful, but cloud watchers categorize EQC in the following ways (Kamei, 1976).

Dragon-like or snake-like clouds and swelling shield-shaped clouds: Also referred to as tornado clouds, snakelike clouds surrounding the sun are said to indicate an imminent severe thunderstorm or large earthquake. In Figure 2.2 the extrapolated direction of the tail of the dragon-like (tornado) cloud indicates the epicenter of the Kobe Earthquake.

Vapor trails: Contrails suddenly form in a blue sky and persist at the same position for a few hours, sometimes turning pink and yellow in the sunset. The epicenter direction is said to be perpendicular to the trails. (See Figure 2.1, also showing earthquake light and the background of Figure 2.2.)

A mackerel sky: Clouds with rainbow fringes.

Concentric trails: The center of the circles is said to indicate the earthquake epicenter.

Radiant clouds: Extending upwards and outwards.

Fan-shaped clouds: The base indicates the epicenter.

Belts of cloud: Sometimes large belts of cloud divide the sky.

Cirrus: A cloud type forming wispy, filamentous threads or streaks at high altitude above a geological trough. The extrapolated direction from the end point of the cirrus cloud denotes the epicenter of a foreseeable earthquake (Sasaki, 1987). Some shapes are also schematically shown in Figure 7.10 (Kagita, 1983).

Clouds similar to those alleged to be EQCs are also formed meteorologically in nature, which makes people sceptical of EQCs. But cloud watchers say there are differences. Firstly, most EQCs form at an altitude of 3-6 km, while rain clouds form below 3 km, secondly EQCs maintain their shape and position for a few hours, while natural clouds drift and change shape according to wind direction (disappearance by diffusion and repeated generation at the same position may cause EQCs closer to the ground to stay stationary), and thirdly, some EQCs glow. Cloud watch-

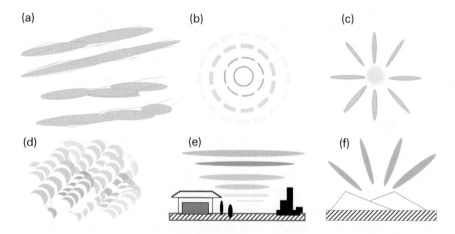

Figure 7.10 Shapes of earthquake clouds (EQCs) as observed from the ground (Kagita, 1983). (a) White belt shaped clouds in a blue sky two days before an earthquake. (b) Concentric trails like wave ripples. (c) Radiant clouds. (d) Cirrus. (e) Horizontal clouds produced by (b). (f) Another depiction of radiant clouds.

ers have attempted to link clouds seen in Japanese skies with earthquakes in China, Mexico, and Turkey or other distant regions, though seismologists and geophysicists argue their claimed predictions are no different from random.

7.5.3 Formation mechanisms of EQFs and EQCs

(a) Charged aerosol

Humboldt made the following observation about an earthquake in Venezuela (1799).

> *Several minutes before the first shock, an intense storm blew up, followed by rain with large drops. I immediately observed electricity in the air with the Volt electrometer... as the most powerful electric discharge took place at 4:12, two earth tremors occurred, 15 seconds apart.*

This could be plausible evidence for measurable voltages accompanying earthquakes, but Tributsch (1982) concluded that charged aerosols were being emitted from the fault zone, and could explain the formation of both EQC and EQF. While EQF formation is not ruled out, aerosols would take a long time to diffuse to several hundred meters height, so they don't account well for EQC. Nor does an aerosol explanation fit the supersonic speeds claimed for some cloud formation (e.g. the Kanto cloud and some tornado clouds). So, aerosol emission offers a partial explanation, but increased voltages and EM waves at ULF—which also generate aerosols—have better explanatory power.

(b) Intense electric field of EM pulses

The dark discharge model showed that dark glow-type discharges (EQL) could be created by the appearance of an intense charge at the hypocenter area. An electric field gradient collects dust, smoke and vapor and helps them coagulate. An increase in atmospheric charge under an intense electric field produces nuclei of water vapor condensation. As a result, charged aerosol may be generated there by an intense electric field and become nuclei of precipitation leading to fogs and clouds in a supercooled atmosphere. Under thunderstorm conditions similar intense electric fields could also produce thunderclouds by condensation.

Vapor trails might also be formed by an electric field, producing a stable pattern of high and low intensity standing waves at long distances from an epicenter. If this is the case, EM waves may also generate clouds at some distance from the epicenter.

7.5.4 Experiments to generate EQFs and EQCs by electric fields

(a) Chamber for cloud formation

Educational experiments on the formation of fogs or clouds were performed under an electric field in a supercooled atmosphere using the classic Wilson's cloud chamber

(Wilson, 1987; Teramoto and Ikeya, 2000). The cloud chamber was made of Lucite (Perspex) and a bottom copper plate was placed over broken pieces of dry ice to obtain a uniformly cooled ground plate as shown in Figure 7.11 (a). A sheet of sponge soaked in ethanol was attached on the inner side of the upper lid of the cloud chamber. Evaporated ethanol passed toward the lower plate forming a supersaturated cool atmosphere over the plate.

Another arrangement shown in (b) indicates a needle electrode set to produce an intense electric field at its tip. The electric field intensity at the site where clouds formed was estimated by measuring the actual voltages at various points using a probe dipped in a separate water bath experiment.

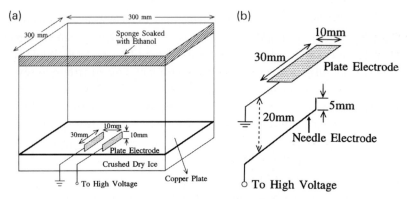

Figure 7.11 (a) A schematic drawing of a cloud chamber which generates clouds by applying a high DC voltage between two plate electrodes and (b) an arrangement of needle electrodes to produce a tornado-like plume (Teramoto and Ikeya, 2000).

(b) Experiments on formation of electric clouds

Formation of clouds in the cool ethanol atmosphere was observed when an increasing voltage was applied to the left plate electrode (anode). In Figure 7.12 (a) at 1.2 kV, a cloud is seen on the anode, with, sometimes, a plume-like "rod" cloud ejected upwards from the edge. With an increase in the applied voltage to 2.4 kV new clouds appeared at the anode as well as the cathode (right electrode). When the voltage reached 3.6 kV, clouds spread in radial direction from the anode along the electric field and attached to the cathode, as in (b). At higher voltages again, clouds from the anode were overtaken by those from the cathode and finally filled the whole chamber as in (c).

The clouds in the area between the electrodes drifted radially along the electric field from the anode to the cathode but the density was higher at the anode, as in (d).

Thin rods (needle-like clouds, also called streamers) perpendicular to the plane of the electrodes were observed on both planes of the anode [also (d)], while they were observed only to one side of the cathode—the anode-side. When the electric field was reversed the behavior of the clouds also reversed.

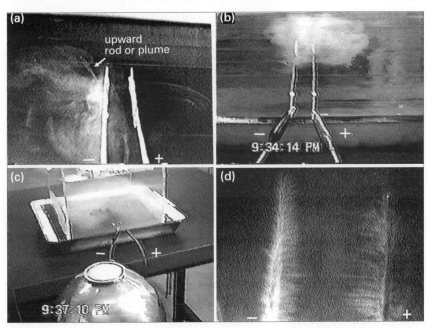

Figure 7.12 Clouds produced in a chamber by an electric field near parallel plate electrodes. (a) Clouds and a small "rod" were produced at the anode at a voltage of 1.2 kV (the clouds are moving earthwards, the "rod" upwards). (b) Clouds spread radially from the anode to the cathode at 3.6 kV. (c) Clouds filling the chamber in another experiment using a Van de Graaff generator. (d) Rod clouds develop along the electric field between the anode (left) and cathode (right), (Teramoto and Ikeya 2000). (Video photos.)

Another experiment in which tin chloride ($SnCl_4$) was exposed to water also filled the cloud chamber with smoke (aerosol). Clouds were formed under the aerosol away from the electrodes at a lower field of 800 V/m.

Explanation:

Electrons produced by natural radiation, are present close to the cathode. Accelerated by the intense electric field they create many charged molecules and aerosol particles. This charged cloud of ethanol droplets moves along the electric field and although they lose momentum by collision, the droplets cause an electric wind or ion wind. The electron avalanche at the right electrode reaches a point at which the electric field intensity is rather high (about 4 kV/m) generating a rod or plume on the left electrode, as shown in Figure 7.12 (a). (Others may be present but not visible.) The rods (streamers) propagate themselves through the atmosphere by continual establishment of an electron avalanche just ahead of their advancing tips as a result of the intense electric field and their own UV-light.

Although the voltages produced in this experiment are higher than likely before earthquakes, discharge from sharp natural points (e.g. pine needles, grass) could raise the voltages to similar intensities giving similar effects.

To return to EQCs, cloud watchers link cloud shapes to future epicenter areas. A fan-shaped cloud, for example, could be generated at an epicenter because of intense charges there and diffuse out into the atmosphere beyond, the same for funnel and concentric EQCs. (The other electrode is very diffuse; earth areas many kilometers away.) Water vapor in the upper strata of the atmosphere is also condensed by the action of charged particles excited by high-energy electric fields which are also capable of producing EQL.

For large inland earthquakes (M>6), retrospective precursor reports say EQCs appear a few days or about eight days before quakes. This links generally with the observation of anomalous peaks in ULF waves before the Loma Prieta Earthquake (Figure 11.4, Fraser-Smith *et al.*, 1990).

7.5.5 Can electric fields produce tornado clouds?

Tornado-type clouds were photographed over the Kobe epicenter zone about 7.5 days before, and again, about 18 hours before, the Kobe Earthquake, as shown in Figure 2.2 (the Akashi Bridge in the foreground) and the color plates. They were observed to maintain their position and shape independent of wind, in spite of a wind flow of 4 - 6 m/s. The clouds appeared up to a height of 2000 meters at what must have been an almost supersonic speed judging from a consecutive series of photographs taken every few seconds. Interestingly, a set of hydrodynamic equations solved numerically assuming a gas jet from the fault could not explain the observation (Enomoto, 2002).

Some have argued that the tornado-shape is formed through a combination of condensation and electric fields.

Experiment:

A needle electrode was connected to the high voltage (240 kV) sphere of a Van de Graaff generator, with the grounded sphere close to the needle electrode. The upper electrode was connected to the ground in the cloud chamber in which the experiment took place. A thin plume like a tornado cloud was swiftly generated at the needle electrode as shown in Figure 7.13, the color plates and the title page of this chapter (Teramoto and Ikeya 2000). The plume disappeared immediately the voltage was reduced to zero. A miniature vortex was also produced from the tip of the needle electrode in the cloud chamber when the ceiling of the chamber was electrically grounded.

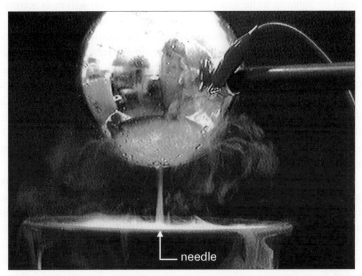

Figure 7.13 An upward stream, like a tornado-cloud, generated by applying 250 kV between a bottom needle and the upper grounded sphere in a supercooled atmosphere (approximating conditions in nature) formed by dry ice and water (See also front color plates).

Explanation:

Cloud particles generated at the tip of the needle electrode by electric charge become further ionized through continuing collisions. An electric wind is produced and a rod or plume-like cloud with an upward convection is produced. A change in the polarity of the Van de Graaff generator from positive to negative did not change this formation. This suggests that both positive and negative aerosols contribute to the cloud formation and that the ion winds cause the plume cloud (Chapter 7, title page) to develop into the streamer of Figure 7.13. A plume between a line source of ions and a flat plate was also studied both theoretically and experimentally. It showed that a tornado-type cloud was formed from a needle electrode on application of a high voltage if an initial vortex movement was formed by nearby fins (McCluskey and Perez, 1992).

Dissipating tornadoes by induced lightning?

Disasters caused by tornadoes in the USA are more frequent than those caused by earthquakes. Though the formation mechanism of a tornado has not yet been established, earthquakes and tornadoes appear to have this in common: they involve electrodynamics. In one USA TV program, a tornado was accompanied by lightning to the ground and sometimes tornado activity is accompanied by other electrical phenomena, such as funnel luminescence.

The tornado has been conceptualized as an electric field generator because it transports charged dust and droplets in a vortex flow (Church *et al.*, 1990). The electric and kinetic energies for the movement of the vortex seem to be strongly coupled so that the electrical energy is converted into kinetic energy through an ion wind effect (Davies-Jones and Golden, 1975). The electric field, and so the ion wind, play a vital role in the persistence of the plume. Thus the stabilization and dissipation of tornadoes would also be related to the electric field in nature; the supply of upward plume energy might be terminated by electric discharges of the electrified clouds.

Scientists in the Institute of Laser Technology, Osaka University succeeded in using a laser beam (for prevention of lightning strikes on power lines) to induce lightning to an electric rod and thus remove atmospheric charge (Yamanaka *et al.*, 1998; Wang *et al.*, 1998). Another method of discharging electrified clouds is to launch a rocket trailing a grounded conductive steel wire. The intense current produced by induced lightning to the rocket heats the wire and evaporates it to plasma. So, artificial lightning discharges of electrified clouds using laser or rocket, as imagined in Figure 7.14, may drain energy from tornadoes and save lives.

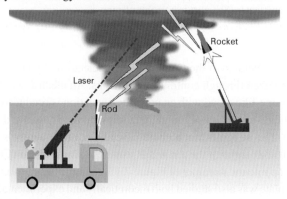

Figure 7.14 A proposal to neutralize a tornado by means of laser-induced lightning and rocket/wire-induced lightning. The lightning discharges the upward plume created by the electric field.

One can speculate that the tornado-type clouds observed before the Kobe Earthquake were precursors. Such clouds may be formed in a supercooled atmosphere above the fault line, especially above the edge of the fault line if an intense charge appeared on the ground. A highly lithic fault gouge (as if baked) was found at the Nojima fault (part of the Kobe Earthquake movements) and argued to be evidence of high current heating by EQL (Enomoto and Zheng, 1998). It may have been formed at the same time the tornado cloud was generated.

Tornado-type clouds have been observed, and no quakes have followed; on other occasions they have. Generation of charges on the ground at a fault zone by local fracture does not necessarily lead to the massive fractures that allow movement of a fault if the movement is then blocked elsewhere. In that respect, many EM macroanomaly phenomena may only indicate large-scale underground fractures which may or may not lead to large earthquakes.

7.6 Unusual phenomena in the sky

7.6.1 Sun, moon and stars

(a) Color of the sky

Old proverbs say,

> *Extraordinarily high temperatures and a red sky are signs of a large earthquake.*

Reports of red skies before earthquakes in Japan and Turkey might come from the scattering of light by an aerosol. Aerosols, mostly of water emanating from the ground would clump together in the sky, leading to a high-density aerosol and change in the appearance of the sky. An extraordinarily red sky in the evening or yellow sky during the daytime (Chapter 1) can be explained as scattering of light by aerosol generated or collected by the intense electric field. Long-lasting red colors could also conceivably be produced by mechanisms discussed in Section 7.4.2.

(b) Sun with halo

Small ice particles formed by condensation in an intense preseismic electric field would form a halo as expressed in the Chilean proverb about showers or rain or an earthquake after the appearance of a haloed sun (See Chapters 1 and 2). The unusual appearance of the sun may result from viewing it through EQCs or through partly condensed water vapor caused by the electric field. Alexander von Humboldt described the sun as "spreading wide, distorted and flared out on its rim" before the earthquake in Venezuela in 1799 (Tributsch, 1982).

(c) Moon with halo and elongated moon

Humboldt described his observation of a moon halo of 12 degrees diameter before the earthquake mentioned in Section 7.5.3. According to Japanese proverbs (Chapter 1) moon haloes indicate rain or an earthquake. Haloes are formed following vapor condensation, and light scattering can give them a red color. The elongated moon seen before the Kobe Earthquake in Figure 2.3 might be due to viewing the moon through the "lens" of a vapor trail, as illustrated in Figure 7.15.

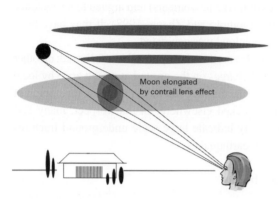

Moon elongated by contrail lens effect

Figure. 7.15 An elongated moon observed before the Kobe Earthquake in Figure 2.3 may be explained by the presence of vapor trails (EQCs) acting as a lens.

(d) Stars twinkling vividly or moving stars with tails

Fluctuations in atmospheric density can make stars twinkle and look larger. The vividly winking and extraordinarily bright stars mentioned in legends and earthquake precursor reports could also be due to a lens effect similar to that described in (b) and (c), in which starlight is intensified by inhomogeneous vapor condensation. A report that stars moved horizontally like falling stars with tails before the Izmit quake indicates the one-directional movement of the condensed vapor which acts as a lens.

Another less plausible explanation is that the sky is "cleaned" just outside the epicenter area by the collection of aerosol to the epicenter by an intense ground charge acting like an atmospheric industrial electrostatic dust collector. This might make stars appear brighter just outside the epicenter area.

(e) Short and blurred rainbow: Evaporated Mukuhira arc

A short arc rainbow or a short vertical rainbow has been associated in Japan with earthquakes or droughts. In different parts of Japan the rainbow has been given different names: Sun child, Fire child, Heaven fire, Fire powder (Daigo, 1985; Musha, 1995). The arc has been called the Mukuhira Arc because of its association with a man called Mukuhira who fraudulently capitalized on the arc legends to forecast earthquakes (Fujiwara, 1937). By the time the fraud was exposed, Mukuhira had involved several Japanese academics who had by then named the arc, the Mukuhira Arc. Consequently, reports of short arc rainbows before quakes have not been believed, in spite of a considerable history of earthquake rainbows before that time.

Explanation:

Rainbows are formed by water droplets. Blurred rainbows could be formed when an inhomogeneous distribution of water droplets diffracts sunlight. A short arc can also be formed when a horizontal cloud or vapor trail is present and acting as a lens, as shown in Figure 7.16. As explained, these clouds may be formed by an electric field or interference of EM waves before earthquakes (Ikeya *et al.*, 1997b).

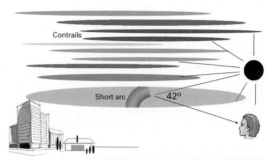

Figure 7.16 A short rainbow called "fire powder" in Japan is said to be a precursor of earthquakes. Intense electric fields of EM waves may form contrails that diffract sunlight and act as a lens.

The appearance of an intense seismic charge on the ground at the epicenter and the formation of lenticular condensation (latent cloud) induced by an electric field in a supercooled atmosphere can explain all the phenomena in this section.

7.6.2 When and where before the Kobe Earthquake?

Section 2.8 showed a relationship between time and reported species behavior before earthquakes, indicating the precursor reports were not merely inventions. The late Professor Y. Mitsuda, also produced a series of tables that showed a space-time relationship for observed meteorological phenomena. His tables on the location and time dependency of reported meteorological anomalies in the Primary Report by the Kansai Science Forum are given in Table 7.2. Mitsuda, a meteorologist who was skeptical of reported phenomena, remarked that reports of condensed water vapor and EQC before the Kobe Earthquake were inconsistent with local weather conditions at the time. The tables indicate differences in time and location (in relation to the epicenter) for sun, moon and clouds e.g. reports of unusual sun at the epicenter on the day, and a day before; of an unusual moon (about a day before) and less than 30 km away from the epicenter; of unusual clouds more than 30 km away from the epicenter and more than two days before the earthquake. The time-location differences in each table are statistically significant ($p < 0.05$, based on statistical treatment by Whitehead *et al.*, 2003) and cannot be attributed solely to psychological factors.

7.6.3 Prediction of the longitude of earthquakes from red aurora?

Charged particles, which stream constantly from the sun as a type of solar wind, are trapped by the earth's magnetic field, and blue auroras (blue light from excitation of

Table 7.2 The number of reports on unusual sun, moon and clouds before the Kobe earthquake for observation sites at a distance R and precursory time before the Kobe earthquake.

(a) Brightness of sun and daytime sky

Place\Time	< 1 d	~ 1 d	> 2d	Total
Epicenter	15	0	2	17
$R < 30$ km	8	14	10	32
$R> 30$ km	1	2	1	4
Total	24	16	13	53

(b) Color of moon and night sky

Place\Time	< 1 d	~ 1 d	> 2d	Total
Epicenter	0	3	2	5
$R < 30$ km	4	24	7	35
$R > 30$ km	9	11	5	25
Total	13	38	14	65

(c) Unusual clouds

Place\Time	< 1d	~ 1 d	> 2d	Total
Epicenter	2	10	8	20
$R < 30$ km	2	28	53	83
$R > 30$ km	0	27	23	50
Total	4	65	84	153

The Kansai Science Forum (1998): *Primary Report on Utilization of Earthquake Precursor Information*, Section 2.4 Evaluation of Meteorological Data by Y. Mitsuda.

oxygen by the charged particles) are observed, mostly near the poles. Dr. S. Uda linked an unusual red aurora in the north polar sky to the Izmit Earthquake in 1999 and thereafter successfully predicted the E/W longitudes of big earthquakes in Taiwan and Indonesia in 1999. His argument is as follows.

An intense electric charge is induced locally by EM waves at ULF before earthquakes, which draws the ionosphere down towards the Earth at that point and scatters FM (VHF) waves, as described in Chapter 11. This bending of the ionosphere, if it occurs, allows the penetration of the charged particles deeper into the Earth's atmosphere at the polar regions following the Earth's magnetic field. The combined effect of this is to alter the color of the aurora (from blue to red) and to protrude the aurora perceptibly along the line of longitude shared with the epicenter, allowing predictions for large earthquakes (east-west) along a meridian of longitude (but not north-south along parallels of latitude).

However, many factors, e.g. magnetic storms from solar flares, affect the behavior of the auroral shape just as local weather affects the shape of clouds in nature. Obviously specialist expertise is required to investigate possible links between auroras and earthquakes.

7.7 Summary

A fault generating an intense electric field and EM pulses can explain unusual phenomena such as earthquake light, clouds and fogs, as well as miscellaneous phenomena in the atmosphere and sky before large earthquakes.

Electrons in the atmosphere are accelerated by the electric field at the fault zone and produce light when they strike atmospheric molecules. Generated UV light also ionizes the atmosphere. The charge distribution and the inhomogeneity of the electric field would contribute to different shaped EQL for different earthquakes.

The EM (dark discharge) model of a fault showed not only that a luminous dome could be produced and be visible one kilometer away even under a full moon, but also that EQL could be seen above the tree level on mountains, since the intense electric field formed by fractures of underground bedrock would extend well above the surface of the ground. It also showed that electric field intensities were higher before a fracture than at fracture itself, lending support to reports of pre-seismic EM phenomena.

Electric charges induced on the surface of the ground may also induce lightning strokes from clouds to the ground or between clouds—consistent with the old saying:

The year of much lightning is the year of many earthquakes.

Earthquake clouds and fogs could be formed during the daytime by vapor condensation due to intense electric fields in a supercooled atmosphere, much as the experimental clouds and fogs formed experimentally at electrodes in a supercooled atmosphere in a Wilson's cloud chamber. A short rainbow might be visible when water droplets created by electric fields diffract sunlight and are viewed through a contrail acting as a lens. Concentric ring clouds may be formed like standing waves above the epicenter by the interference of scattered EM waves.

The old folk stories and proverbs on light, clouds and fogs before earthquakes may, thus, have some scientific basis. Although these phenomena are only fringe subjects in orthodox meteorology and anecdotes in seismology, they may yet become a substream in Electroseismology with some predictive usefulness.

8

Precursor Phenomena

On Land, Sea and Elsewhere

One of the more unusual earthquake reports was from a Turkish fisherman who said his boat was dragged to the seabed between walls of water during the 1999 Izmit Earthquake; an event we have dubbed "The Moses Phenomenon". This chapter offers a mechanism for this unusual report, and an assortment of other reported unusual earthquake precursor phenomena. Above: Moses leading the Children of Israel over the Red Sea (by Nami).

8.1 Introduction

Unusual earthquake precursor signs are not limited to the biosphere and the atmosphere. Geophysics textbooks have described land tilting, rising water levels in wells and increased temperatures of hot spring water before earthquakes (Chapter 3). Some legends talk of increased cloudiness of water, and often of tidal withdrawal; others of low frequency sounds. Many individuals reported hearing peculiar low frequency sounds from the ground before aftershocks of the 1995 Kobe Earthquake. Musha and Rikitake document reports of explosive pre-quake sounds from underground and from mountains, and such sounds were also heard before and after the Kobe quake.

This chapter argues some of these phenomena can be explained by the generation of EM waves: in particular reports of some low frequency sounds and water turbidity. It also proposes EM effects for other reported pre-quake anomalies: behaviors of candle flames and smoke, a malfunctioning thermometer and the unusual movement of an air bubble in a surveying level.

The chapter also proposes a theory to explain the unusual tale of a Turkish fisherman during the 1999 Izmit Earthquake. The man said his boat was borne down between walls of water to touch the floor of the Gulf of Izmit, before being carried towards land by big waves (Chapter 2).

The unusual behavior of a magnet and domestic electric appliances, which are also explicable as EM phenomena, are separately discussed in the next chapter.

8.2 Air bubble movement—from land-tilting or electric charge?

8.2.1 Movements of air bubble in a level before an earthquake

A level with an air bubble in a slightly bent glass tube is a standard piece of equipment in geodetic land survey, used to determine the horizontal plane. In 1944, military engineers surveying near the shoreline in Shizuoka prefecture in Japan were unable to carry out measurements because of violent movements of the air bubble in the level both two days before, and two or three minutes before, the Eastern Nankai Earthquake (M7.9), whose epicenter was 170 km away. The odd behavior was ascribed to a rapid movement of land, a rebound of the subducted plate. However, no seismic waves were recorded at the time the air bubble moved so violently.

8.2.2 Could it have been an EM phenomenon instead?

The level was manufactured just before the end of World War II when there was an extreme metal shortage. It was made of chestnut wood fitted with a bent glass tube and a brass plate. In other words, it might not have been shielded well electrostatically.

Experiment:

A plastic level was placed on an acrylic plate close to the Van de Graaff generator and the electric field was rapidly turned on and off by touching the high voltage sphere with the grounded metal sphere. The electric field changed greatly as the grounded sphere was moved. The air bubble in the level moved back and forth as in the photograph in Figure 8.1 (a) and (b). Though some would argue that such an effect was inevitable given such a high voltage (250 kV), a plastic sheet charged by mechanical rubbing and moved close to the air bubble had a similar effect.

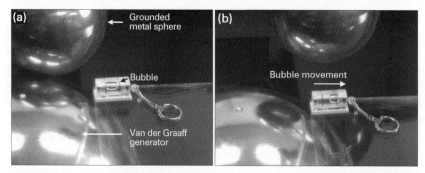

Figure 8.1 An experiment showing the movement of an air bubble in a level in a changing electric field. When the grounded metal sphere was moved from (a) to (b) near a high voltage Van de Graaff generator, the bubble moved to and fro. Possibly, like the level used by the engineers, it was not electrostatically well-shielded.

Geophysicists strongly object to such an explanation, and reasonably so if they are speaking of modern electrostatically well-shielded equipment. But, as mentioned, this was probably not the case in 1944. A senior seismologist remarked that there was no record at the time of a seismic signal in a seismograph close to the area where the surveying was being done, and that the air bubble's violent movements might be attributable to EM pulses.

Professor Rikitake, however, commented that the base line of the seismograph was too noisy to rule out the possibility of land tilting as a cause and that all phenomena should not be forced into an EM model. The author was not aware of such seismograph data but still maintains that intense EM effects are a plausible explanation, though the possibility of slow land tilting, which produced no clear signal in the local seismograph, cannot be excluded. Land movements before and after recent earthquakes e.g. the Taiwan-921 Earthquake, measured by satellite GPS have not yet reported significant presesimic ground tilting. This offers some support to the possibility of EM effects.

Pendulum swing anomalies—an electric field effect?

The Nobel Prize Winner, M. Allias, observed remarkable disturbances in the movement of a Foucault Pendulum (which measures the Earth's gravity), at the time of solar eclipses in 1954 and 1958 (Allias, 1959). A torsion pendulum also showed anomalies during solar eclipses in 1970, on June 30, 1984, and on August 11, 1999. However, the effect of gravity on the pendulum as a result of solar or lunar action is 100 million times too small to influence a pendulum, and in the 1999 eclipse the change of a few degrees in the azimuth (outer swing) of the plane of the pendulum was ascribed to atmospheric pressure modulation caused by the moon's shadow.

However these anomalies might be explained as electric field phenomena caused by a solar eclipse, in that an ascending ionosphere during an eclipse results in changes (a reduction) in the vertical electric field, which, then, affects the gravity measurement. Both the change in the azimuth of the pendulum and the vigorous movement of the air bubble in the level, possibly caused by charges induced by an electric field at ULF (See Section 8.2), could be a result of change in electrostatic induction.

A simple way to test this is to observe the swing of a pendulum subjected to differing electric fields. In a laboratory test we found that the frequency and azimuth of a metal pendulum changed under external electric fields in response to electrostatic attraction between the charges on the metal pendulum and a ground aluminum plate connected to a Van de Graaff generator.

Gravity anomalies have been reported to have been observed before earthquakes. In the author's view, underground dilatancy formation and fluid movements producing electric fields of EM waves may be a factor in this.

Electric charges in thunder clouds might also change the period and azimuth of a pendulum, especially one of light metal.

It would be of interest to monitor pendulum movement to observe any correlation with lightning and earthquakes, though the electric forces themselves might affect the measured gravity.

8.3 Sounds from the ground: Can one hear EM waves?

8.3.1 Records of strange sounds and sounds from the ground

Many people reported hearing low frequency sounds just before the Kobe Earthquake and its aftershocks— before they felt any tremors. In fact some people claim that earthquake sounds just before a tremor alert them that one is about to occur. The author, in Osaka—40 km away from the epicenter—heard nothing before the Kobe mainshock, but many did. New Zealanders say they often hear sounds just before an earthquake.

On the face of it this is puzzling because the speed of sound in air, 340 m/s, is slower than that of seismic P- and S-waves: 8 km/s and 4 km/s respectively. People cannot be hearing sounds generated by P-waves, when the sound could not yet have arrived.

The Russian writer, P. Skitaretzb, recorded his own experience of sound before a quake, in the Bulletin of the Seismological Society of America, Jaggar (1923).

We walked in the middle of the road, followed by the man pulling the cart. Then suddenly, very near the bluff, I thought I heard the sound of an approaching train. I was surprised, for I know no train was near there. I was about to mention it to my wife, but there was no time. From somewhere I heard the roaring of a wild animal, and a sudden fierce wind came up and bent the branches of the trees like a bow. A sound like an underground train came from the direction under our feet, seemingly some pent-up, awful energy, seeking escape. The angry roaring increased; enraged shaking was coming upon us. The ground began to move, groaning and yanking us back and forth with the mad speed and fanatic energy of a lunatic. We felt as though we were about to torn into pieces, and great Earth was trying to shake off everything on it.

A young seismologist, who believed he had heard low frequency sounds before a Kobe aftershock, immediately observed his seismograph—which was operating at the time. There was no signal. Then, only moments later it began to record the P-wave and tremor (S-waves). When he told his colleagues they were derisive and suggested he shouldn't circulate the story if he wanted to remain on-side with his professor. His account inspired the following experiment.

8.3.2 Sound and converse-piezoelectricity

If there is an intense electric field, piezoelectric materials shrink or elongate under an effect which is the reverse process of piezoelectricity: converse-piezoelectricity. (That is, electricity is converted to stress rather than stress to electricity.) If piezoelectric quartz grains in granite are struck by stress-induced EM waves, those EM waves will (in converse-piezoelectricity) produce stress in the quartz grains, and sound. Because EM waves travel at the speed of light, coseismic EM waves that have reached the observer might produce sounds heard earlier than sounds generated by the arrival of slower seismic P- and S-waves.

Experiment:

Aluminum tape electrodes, 15 cm x 15 cm, were attached to a granite plate, 40 cm long, 90 cm wide and 2 mm thick (Figure 8.2). Plates of limestone and glass of similar dimensions were also used for comparison. An AC voltage of about 10 kHz was applied to the electrodes to generate an electric field. Sound was generated from the conversion of the electric field into induced stresses on the quartz grains in the granite. The production of the sound in itself is clear evidence of the "converse-piezoelectric" effect at work.

To evaluate the intensity of the sound (hoping this might ultimately allow an estimate of the intensity of the EM waves in earthquakes) a microphone was used to detect the sounds, but it also picked up the intense EM field and generated extraneous sounds. So an aluminum box was used to shield the measuring system and ensure only sound generated by the EM waves in the rock registered. A 10 kHz EM field was used because this was just the right wavelength to interact strongly (resonate) with the dimensions of the rock. EM waves at 10 kHz have a wavelength of

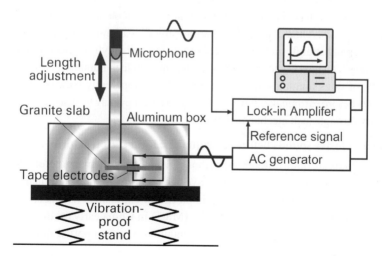

Figure 8.2 Sounds were generated by EM waves in granite from the generation of stress by an electric field (Matsuda *et al.*, 2003).

40 cm and by the laws of acoustics will interact with the 20 cm granite dimension, but had to be slightly tuned to do so. The experiment showed that the intensity and type of the generated sounds depended, as expected, on the dimensions of the granite and its various possible vibration modes, and that the precise mode at 10 kHz was one of the "bending vibrations" it possessed. The sound originated from the many quartz grains and could not be heard from glass or marble, which do not contain them.

In this converse piezoelectricity, under a voltage of 10 kHz, the quartz grains shrank and expanded 10,000 times a second producing a sound which was also 10 kHz and had a very high pitched rather pure tone. The situation would be much more complex in a real earthquake because the frequency range and varying pulse widths of the EM waves produced would be capable of producing complex or low frequency sounds well ahead of P- or S- waves.

8.3.3 Can animals directly hear EM pulses as sounds?

Is it possible that animals and some humans can actually detect EM waves as sound?

The author used to work in an room adjacent to a student, Dr H. Matsumoto, who insisted he could tell when the author's computer was turned on every morning and when the monitor switched from sleep to functional mode. We wondered if he might be detecting EM waves but Matsumoto said he was hearing sounds. The author could hear nothing, but, typically for a man of his age, is unable to hear frequencies above 6 kHz. However Matsumoto had an exceptional range of hearing—up to 20 kHz, as laboratory tests showed.

He may also have been detecting EM noises as acoustic waves through small hairs in his ears. A hair cell is a "mechanoreceptor"—a sensory organ that responds to mechanical stimuli such as pressure or tension and maybe to electric fields. Because dogs can hear higher frequency sounds inaudible to humans it is possible that reports of dogs digging furiously in the soil and apparently barking at invisible enemies before earthquakes have a scientific basis: they may be detecting EM fields directly through such mechanoreceptors.

8.3.4 Explosive sounds before earthquakes: local rock fracture

These sounds are quite different from those arriving only moments before P-waves. They are mentioned in literature and have been reported before earthquakes. They have often been said to sound like cannon fire or explosions with accompanying echoes. Before the Haiyuan Earthquake (M8.5) in China early on the evening of December 6, 1920, people living in a mountainous region reported strange sounds echoing in the valley the midnight before.

During the rock crushing experiment described in Chapter 5 the partial fractures of rocks of 20 cm x 20 cm x 40 cm by a 1000-ton press sounded like explosions; they made the laboratory floor tremble.

The author also believes he heard such sounds emanating from an area well known as the source of the Inagawa Earthquake swarms that occurred before the Kobe Earthquake. (These have been considered foreshocks by some but have also been attributed to stress created by small dam constructed in the locality ten years before.) The sounds occurred two months after the Kobe Earthquake: strange intense beating sounds—as if someone were hammering a rock or pounding a ball against a wall. It being 5 a.m. the author rose to investigate the sound, or at least to protest. But the sounds did not come from the immediate vicinity. Instead they emanated from the Inagawa swarm area at the edge of the town, where they sounded like rifle-fire, or like rocks being broken up at a quarry.

Such sounds may be generated by local fractures in massive rocks blocking fault movement. Although EM pulses are generated by such fractures, the fractures themselves are sufficient to explain the sounds.

8.4 Phenomena in the aquasphere

8.4.1 Turbid water in wells and bottles

(a) Turbidity caused by electric current in water

There are reports of white turbid (cloudy) well or spring water before earthquakes (Rikitake, 2001; Wadatsumi, 1995).

> *The well water suddenly became turbid and then the earthquake came.*
> *The seafloor of the Seto Inland Sea became turbid before the Kobe earth*
> *quake and also before its aftershocks (NHK).*

Small earthquakes or microcrack formation may lead to the suspension of tiny rock particles in underground water. An intense electric current in water might be another cause of turbidity, as demonstrated in an experiment applying voltage to a needle electrode in a bottle; the electrolytic precipitation of dissolved minerals produced cloudiness. Well water may be thus electrolyzed by the presence of iron chains or wet ropes acting as electrodes. Small bubbles of hydrogen and oxygen and resuspended mud may also be created by current flow in water.

(b) Does wine become turbid and lose flavor on exposure to an intense electric field?

There was one report of mineral water in a bottle becoming turbid before a quake due to precipitation of minerals, and several from Italy claiming wine turned turbid and unpleasant to the taste before quakes. To the delight of the graduate students a laboratory experiment was conducted on bottled wine and mineral water—and at their insistence bottled beer (since canned beer is electrically shielded). Unfortunately the samples could not be purchased from the research budget!

Electromagnetic pulses were generated by electric discharges and applied to cups of different wines. No changes were detectable. However, if current was induced by applying a few volts, the wine became turbid from the production of small bubbles of hydrogen and oxygen. However without current flow, there can be no electrolysis. One can imagine a chain or wire acting as an electrode in the case of well water, but in bottled wine there are no electrodes to produce hydrogen and oxygen bubbles. So reports of Italian wine turning turbid before earthquakes cannot be explained as an EM phenomenon without more clarification e.g. the wine's being re-corked after contact with a metal cork-remover.

8.4.2 A split sea in the Gulf of Izmit: "The Moses Phenomenon"

A statement influenced by the Koran?

During the field investigation of earthquake precursors in Turkey a fisherman told an extraordinary story in which his fishing boat was momentarily pulled to the seabed between walls of water during the Izmit Earthquake (Section 2.3.6).

Was the whole account merely an invention or exaggeration by a publicity-seeking Moslem fisherman who knew the Koranic story of Moses crossing the Red Sea?

> *Then we revealed to Musa [Moses] 'Strike the sea with your staff.' So it had cloven asunder, and each part was like a huge mound.* (The Poets 26:63)

However there are parallel earthquake reports from elsewhere so it seems unfair to discount the story on the basis of similarities to the Koranic account.

Tributsch (1982) referred to another like it from the 1783 Calabrian Earthquake. Two peasants working on a high bank felt a tremor and saw the sea split into two parts. During our Turkish fieldwork an independent eyewitness from the Western edge of Istanbul described a dramatic north/south splitting of Küçükçekmece Lake, from which a stroke of lightning ascended.

So the fisherman's story could just be one of a number of reported hydrological phenomena accompanying earthquakes e.g. the legendary account of a reversal of flow in the Seta River (near Kyoto) before an earthquake, and a receding ocean and exposed sea floor just after an earthquake (Chapter 1, Living God).

It is not proposed to explain the above accounts (except perhaps the lightning from the lake) by high electric charge. The "Moses Effect" is the name that has been given to the bending of water levels by an intense magnetic field, of the order of Tesla (T). However, as already explained, the change in the magnetic field before earthquakes are orders of nanoTesla, so it is impossible "The Moses' Phenomenon" could be an EM effect. Instead we suggest an alternative hydrological mechanism related to the fault itself.

8.4.3 Hypothesis: Water movement into faults and areas of subsidence

The hypothesis is that water movement into faults can at times be so intense that it can create momentary "walls" as the surface water disappears. We dub this "The "Moses Phenomenon".

Theory:

Water is known to be absorbed into faults during fault movement. This became clear during drilling of the Nojima fault after the Kobe quake. (The movement of the Nojima fault caused the Kobe Earthquake). At a depth of 1800 m cracks were filled with clay (a relic of water penetration) and there was evidence of oxidized material (from infiltration of surface water deep into the active fault), (Uda *et al.*, 2001). Less directly, in the same fault, the presence of water-filled fractured rock was demonstrated by subtle changes in the velocities of Kobe Earthquake after-shock P- and S- waves through the fault zone (Zhao *et al.*, 1996).

In Turkey, the North Anatolian Fault cuts into the Gulf of Izmit (See Figure 2.9) for about 10 km. The Gulf is partially blocked off from the Sea of Marmara by Cape Golcuk. We assume, as typical for so mature a fault, that it has a fracture zone 100 m across. Its depth is also about 10 km. As described in many freshman texts a typical porosity for unfractured rock is 0 - 5%, but fault porosity in fresh faults is typically 30% (Soler, 2000). The volume of water within the Gulf of Izmit east of Cape Golcuk, is about 10,000 m x 5000 m x 10 m, which equals 5×10^8 m^3 (500 M cubic meters). If 30% of the undersea fault surface east of Cape Golcuk is permeable to water that would be an area 10,000 m x 100 m x 3×10^{-1}, which equals 3×10^5 m^2 (300,000 sq. meters) The main movement of an earthquake (hence water wall) often continues for 300 seconds. If the rate of "pumping" (sucking) by the fault is that of gravity, equivalent to 10 m sec^{-1}, the water absorbed in 300 seconds would be 300 s x 3×10^5 m^2 x 10 m sec^{-1}, which equals 9×10^8 m^3 (900 M cubic meters).

This means that strong hydrological effects due to water absorption could be possible within a reasonable range of estimated starting numerical values. This calculation does not even consider the volumes of water drawn off into areas where land has subsided (which were significant near Izmit), nor does it consider that horizontal flow along the faultline could occur.

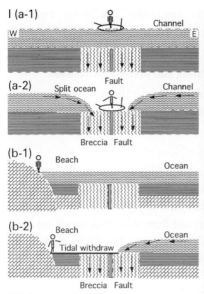

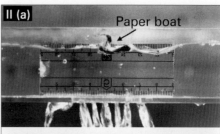

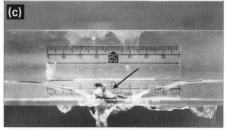

Model experiment:

(i) A narrow channel hydraulic experiment

In Figures 8.3 I (a-1 and a-2) and II [(a)-(c)] the hypothesis is shown diagrammatically and in photos. When water is rapidly withdrawn from the tank the fisherman in his paper boat rapidly sinks to the plastic grating at the bottom (equivalent to a porosity of 50%). Walls of water are on either side of him. The curved nature of these was reproduced in a theoretical calculation (Ikeya *et al*,. 2002). If the grating was suddenly blocked, waves formed and moved away from the area, which accords with the fisherman's description: "then big waves bore me towards the Turkish Naval School building".

Figure 8.3 I (a-1) & (a-2) A schematic drawing of "The Moses Phenomenon" based on the geography around Cape Golcuk where the event occurred. I (b-1) & (b-2) A diagrammatic representation of what might be a secondary process in reported ocean withdrawal and seabed exposure after earthquakes. II (a-c) Photos of an experiment demonstrating "The Moses Phenomenon". The sea "splits" in a narrow channel (the "Gulf of Izmit" when water discharges through a grating (into a "fault zone") and inflow is limited. Walls form and the fisherman and his boat are drawn down between them.

(ii) Model of the Gulf of Izmit

In another experiment a model was made of the Gulf of the Izmit around Cape Golcuk (See Figure 8.4), horizontal scale 3:100,000; vertical scale 1:200. Water

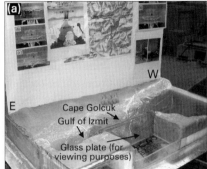

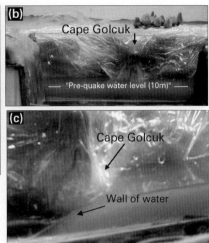

Figure 8.4 (a) A model of the Gulf of Izmit round Cape Golcuk where the fisherman's boat touched the seabed. (b) The pre-quake water level of "10 m". (c) The wall of water produced at "Cape Golcuk" after the "fault" had swallowed water but rapid inflow was blocked by the Cape.

was drained out near the Cape area but inflow was blocked by the Cape and a wall of water formed there. If draining was stopped, a series of waves spread from the area, as they did in Experiment (i).

We also collected a statement that waves flooded a tea garden and a park on the coast of the Gulf of Izmit on August 11, six days before the earthquake. The local people have seen these often enough that they call them "death waves" and consider them precursors of a large earthquake. Such waves could arise from momentary, pulse-like absorption of water by growing microcracks within the fault before its main movement. Legends of death waves before earthquakes in the area may have this physical explanation.

8.4.4 Tidal withdrawal

Figure 8.3 I (b-1) and (b-2) show that if a fault crosses a beach it could absorb seawater, adding to the tidal withdrawal known to occur before tsunamis. It must be further hypothesized that any cracks would reseal and not be visible to observers when the water is withdrawn by whatever means, and that water supply from the ocean is limited by geographical features or characteristics of the earthquake.

8.5 Bent candle flame, smoke and color of steamed rice

8.5.1 Candle-flame phenomena: Bent or extinguished flames

A Japanese proverb on candle flames goes,

Candle flames on temple altars bend like archery bows before earthquakes

One precursor statement related to the Kobe Earthquake mentions the difficulty a religious man had lighting a candle on a Buddhist altar one hour before the earthquake (Wadatsumi, 1995). It was as if the match flame avoided the candle, he said. He mentioned the mystery to his wife and said that an earthquake might be coming. A report from China mentions that a steadily burning candle flame suddenly went out before the Tangshan Earthquake.

Could such statements, and the legend, be explained as an electric field effect and reproduced by an experiment?

Experiments on electric field effects:

The flame of a candle placed on the sphere of a Van de Graaff generator, as shown in Figure 8.5 (a), changed its shape upon charging of the sphere, as in (b). Subsequently the downward-pointing flame melted the candle (producing a flow of red wax over the metal sphere; the ghoulish effect never failed to impress audiences in demonstration lectures).

If a grounded wire was brought close to the flame, the flame 'danced' with the changes in the electric field and then went out.

Flame as plasma:

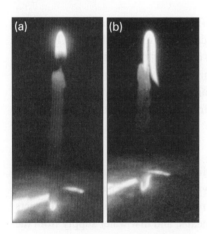

Figure 8.5 Candle flames on temple or shrine altars proverbially bend like archery bows before earthquakes. The same effect was reproduced on the electrically-charged sphere of a Van de Graaff generator. (a) before and (b) after charge was applied to the sphere.

The physics of flames has been investigated in the hope that it would lead to more efficient fuel combustion in motors (Payne and Weinberg, 1958). A high uniform electric field produced by parallel plate electrodes produced a tilt in a candle flame toward the negative electrode.

A negatively charged sphere of a Van de Graaff generator attracts positive ions in a candle flame plasma. Although flame plasma normally moves upward in the heated convection stream, it is pulled down towards the sphere as the heavy positive ions are attracted and pull all other components of the flame down with them, including low-density negatively charged components. Any slight fluctuation in the electric field produces a wavering in the convection stream so that any charged particles escaping the convection are pulled down in an arch towards the charged generator, as in Figure 8.5 (b), dragging other plasma components with them (Ikeya *et al.*, 1998).

However, when the polarities were reversed on the Van de Graaff generator, so that the sphere was positively charged, the flame also bent down towards the sphere because of the high field intensity. So, the above argument is not necessarily true. Perhaps, instead, plasma is polarized by the electric field of the Van der Graaff and moves towards the intense electric field on its sphere, regardless of its initial charge.

A similar mechanism could apply to candles near the ground before an earthquake if the ground has acquired a pulsed electric charge. If this is so, the candle flame will not be stable, not only producing an arch, but making it very difficult to keep alight (Ikeya and Matsumoto, 1997)—as the religious man found.

A report that a plastic sheet crackled just before the Kobe Earthquake supports the hypothesis of appearance of charge, because charged plastic sheets crackle.

8.5.2 Unusual color of steamed rice: Confectioner in Tokyo

A confectioner in Tokyo was proud of his success at earthquake forecasting based on the color of his steamed rice (Rikitake, 1976). On the face it this sounds ridiculous—good evidence for the view that so many small earthquakes occur in Tokyo that anything could be a forecast. However the man was something of a connoisseur of the subtle colors of his steamed rice because he gained his livelihood from making it most days. At the time he made his prediction there were no electric rice cookers or microwave ovens. He would have used wood-fuel and it is not at all unlikely that the flame flickered unstably beneath the rice (because of an electric field on the ground), changing the usual combustion efficiency and producing a subtle change in the color of the rice that only a master of his craft would notice.

According to a record in Rikitake's database (1992), a woman could not get rice to boil six hours before the Niigata Earthquake (M7.5) in 1964. Again this may be caused by inefficient combustion of wood-fuel. Rikitake's habit of recording even apparently ridiculous stories as data has been very helpful in this present exercise, and will be in future studies, although he was castigated by Tributsch (1982) for his skeptical remarks.

8.5.3 Serpentine smoke from incense and chimneys

Proverbs about incense and smoke before earthquakes say,

> *Incense smoke changes shape on the altars of temples and shrines.*
> *Smoke from a chimney flows downward and writhes like a snake.*

A citizen's report to the Kansai Science Forum said that smoke from a chimney bent like a snake before the Kobe Earthquake.

Experiment:

Burning incense was placed on the high voltage sphere of a Van de Graaff genera-tor. The incense smoke was attracted to the charges on the sphere. Pulsing the charge by alternately grounding and charging the sphere created a serpentine movement of the incense smoke even through there was a horizontal flow of air. (Make sure this experiment is conducted in a well-ventilated room or prepare for a sore throat.)

Explanation:

Smoke is essentially dust, or an aerosol, that can be given a slight electrical charge and be attracted to an intense electric field—like the candle flame. The appearance of pulsed charge on the ground under a wind of constant speed would induce charges in the smoke. The pulsed attraction would produce sinuous, "snake-like" flows of smoke from a chimney.

[Aerosols themselves need not necessarily be charged but can be polarized by electric fields—a property that is used industrially to reduce emission of suspended particulate material from chimneys; the device for installation in chimneys (essen-tially two electrodes installed opposite each other at different heights in chimneys) is called an electrostatic smoke-dust collector and collects the smoke for later dis-posal as a solid.]

8.6 Thermometer: the contact angle of mercury

8.6.1 Thermometer with a rapidly falling mercury level

A man with a high fever for three days before the Kobe Earthquake measured his temperature frequently with a mercury thermometer. He was able to measure his tem-perature, but the mercury then fell very rapidly and did not remain at the high point as thermometers are designed to do. He thought that the thermometer was broken but after the Kobe Earthquake the thermometer again functioned normally.

Skeptics will say it is quite likely that one person in eight million in the Kobe-Osaka area might have flu and a dysfunctional thermometer before the earthquake, and turn the story into another silly anecdote. But it's not so silly—as a physics experiment shows.

8.6.2 Control of contact angle by electric field: Unusual surface phenomenon?

Experiment: the fall of mercury under an electric field

The application of an electric field to a thermometer reproduced an abrupt fall in the mercury level. Even if the thermometer was held with the reservoir upwards the mercury returned to the reservoir without holding the maximum temperature.

Explanation: Electric attraction and an unknown surface property

Consider the structure of a mercury thermometer or a maximum temperature thermometer. The mercury level is intended to indicate the highest temperature recorded
after it has been taken from the patient, and it does so by breaking the column of
liquid mercury at the neck of the glass stem, as in Figure 8.6 (a).

However, if the separated mercury is exposed to an intense electric field, the
two separated sections of electrically charged liquid mercury will be attracted and
will flow back together. In addition, the contact angle of the mercury with the glass
as shown in Figure 8.6 (b) is changed by the presence of the electric field in much
the same way Petrenko (1998) observed for the contact angle between mercury and
ice. The reduced area of mercury in contact with the glass reduces friction, enhancing sliding.

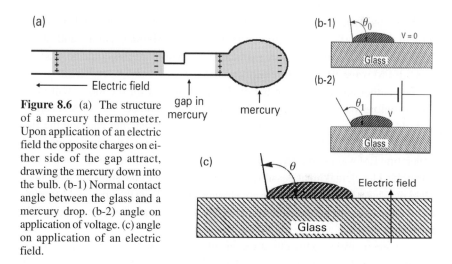

Figure 8.6 (a) The structure
of a mercury thermometer.
Upon application of an electric
field the opposite charges on either side of the gap attract,
drawing the mercury down into
the bulb. (b-1) Normal contact
angle between the glass and a
mercury drop. (b-2) angle on
application of voltage. (c) angle
on application of an electric
field.

The puzzling feature about the mercury after exposure to the intense electric
field was that the altered property of the mercury lasted for a few days, recovering
only gradually, as reported by the lay citizen. A possible explanation is that under
normal conditions the surface of glass (silica) is relatively hydrophobic [O-Si-OH]
meaning the mercury will tend to form clumps and break easily at the neck of the
stem. But an intense electric field will create surface oxygen anions [O-Si-O-],
altering the nature of the glass-mercury contact, and in effect making it much more
slippery. This effect may persist for a few days.

So, another unusual incident before an earthquake may be explained scientifically, and the science may also be useful in technology in which contact angles need altering.

8.7 Summary

Apart from "The Moses Phenomenon" and explosive sounds well before earthquakes, for which mechanical explanations have been offered, the precursor phenomena discussed in this chapter have been reproduced in laboratory experiments using electric fields.

"The Moses Phenomenon" is proposed as a hydraulic phenomenon created by the balance between seawater draining into fault fissures and subsidence areas and inflow limited by obstructing capes and other topography—or in the case of ocean withdrawal—submarine dunes.

The movement of an air bubble in a level, low-frequency sounds from underground, bent candle flames, serpentine incense and chimney smoke and rapidly falling mercury thermometers are explicable as phenomena that could be produced by ULF EM wave effects before earthquakes.

The author appreciates that the explanation offered for the movement of the bubble before the 1944 Eastern Nankai Earthquake could be viewed by seismologists as the ignorant conclusion an interdisciplinary researcher entering the field might be expected to draw. However, the fact still remains, a laboratory experiment generating EM waves produced energetic to and fro movements in the bubble of an unshielded level, making it possible that the movement of the surveyor's bubble in 1944 may have been caused by the appearance of charge on the ground. More clarification awaits further observation of SEMS and of land movements using GPS instruments.

Precursor stories, which may sound ridiculous at first hearing, can prove on investigation to be scientifically verifiable as possible rare phenomena before earthquakes.

9

Mysteries Before Earthquakes

The Behavior of Electric Appliances

Second hands of clocks were reported to have rotated rapidly before the Kobe, Izmit and Taiwan-921 earthquakes and also to have moved backwards. Obviously relativity theory is not a factor! Some have suggested that the *Alice in Wonderland Syndrome*—a vision disturbance disorder—might be responsible, but EM waves may account for the phenomenon. Above: An illustration of the Mad Tea Party by John Tenniel in *Alice's Adventure in Wonderland,* by Lewis Carroll.

9.1 Introduction

The electronic age has added a new category of earthquake precursor reports to the ancient legends.

They are accounts of odd behavior and noises from domestic electric appliances: TVs, radios, clocks, refrigerators, mobile phones, fluorescent lamps, car navigators, and possibly computers. The reports came in independently from Kobe, Izmit, and Taiwan before the large earthquakes of 1995, and 1999 and are described in more detail in Chapter 2.

Interestingly, exposure of some of these appliances to EM pulses in a laboratory reproduced the behaviors described in the reports: clocks stopped, or their hands rotated rapidly; fluorescent lamps lit up, radios went dead, or produced static; color shifts and speckle noise appeared on TV screens, or sets fluctuated between channels; refrigerator compressors switched on and off producing odd sounds; cellular phones illuminated and rang but no record of any call was left; the needle on a magnetic compass fluctuated.

Reports of malfunctioning devices that we did not attempt to replicate in the laboratory setting could just as easily be ascribed to EM wave interference: random self-operation of power windows in cars, spontaneous switching on of air conditioners and intercoms at midnight (five or six hours before the Kobe quake); the apparent sudden switching on of a tape-recorder in Izmit so that a unscheduled call to prayer resonated from a local mosque at 2 a.m.—one hour before the earthquake.

Several reports from slightly earlier than the electronic era are also investigated in this chapter: How was it that nails hanging for weeks from a permanent magnet dropped off two hours before the Ansei-Edo (Tokyo) Earthquake (M6.9) in 1855? Why did iron chains in a military factory begin to swing against each other two hours before the Eastern Nankai (M8) Earthquake in 1944? Why did arc discharges occur between iron bars lying on the ground a day or two before the Tangshan Earthquake in 1976? Why could a man no longer read the phosphorescent numbers on his watch 2-3 days before the Great Kanto Earthquake, but see them again one day later?

9.2 Magnetic earthquake precursors?

9.2.1 Nails falling from a magnet, not a magnetic anomaly!

In 1855 the *Ansei Chronicle* carried a report of 15 cm iron nails that dropped from a big natural magnet two hours before the Ansei-Edo Earthquake. They had been hanging end-to-end from the magnet in a spectacle store for some time as part of an advertisement. The magnet regained its iron-attachment properties after the earthquake.

Attempts have been made to link the anomaly to a change in the earth's magnetic field—but as discussed earlier, the variation in the earth's magnetic field before earthquakes is minuscule; the earth's magnetic field is 0.03 - 0.05 mT, and the change, a few nano-Tesla (nT) or about one ten thousandth of the field (Rikitake, 1986). The surface intensity of a ferrite magnet is tens of mT and that of the most intense permanent magnet made of Neomax (Nd-B-Fe alloy) is about 800 mT. The magnetic field of the natural magnet mentioned in the *Ansei Chronicle* may have been several mT at its surface. The only way to explain the phenomenon as a magnetic one was to argue that a large magnetic field variation had occurred before the earthquake. This was not credible, so the story came to be regarded by most geophysicists as a misleading anecdote, though there were two similar European reports (Rikitake, 2000).

Attempts to link the interrupted magnetic contact to the arrival of seismic P-waves does not stand up either, as they dropped off two hours before the quake. Had they fallen 20 seconds beforehand, the effect could plausibly have been argued to be a P-wave effect, since the epicenter was about 160 km away from the store. However, the story may be explicable as an electric effect.

Electric discharge experiment:

Several nails were attached to a magnet hung from a pole made of plastic LEGO blocks (See Figure 9.1). The floor under the LEGO construction was covered with aluminum foil, which was connected to the high voltage sphere of a Van de Graaff generator. When the high-voltage Van de Graaff sphere was electrically charged, the nails repelled each other. When the voltage was turned on and off (to simulate EM pulses of ULF waves) the iron nails began to swing increasingly, and finally dropped off.

Explanation:

Electric induction by EM pulses:

The iron magnet and nails acted as an electrode when the aluminum foil was charged up. Electrostatic induction generated a charge on the nails and an attractive force was formed between the ground and the nails. The nails, charged at their tips with the same electrical charge, repelled each other. When the charge was removed from the floor, the nails returned to their original position. When it was applied and removed several times, the nails began to swing, became unstable and dropped off.

So the nails that fell from the magnet at the spectacle store two hours before the Ansei Earthquake may have been responding to electric charge appearing on the ground, and not to changes in the Earth's magnetic field or to land tilting. This transforms the tale from a magnetic anomaly to an electrical one.

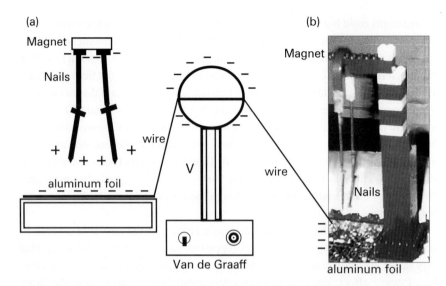

Figure 9.1 The induction of opposite charge at the nail tips made them drop off a magnet—as they did two hours before the Tokyo Earthquake (M6.9) in 1855. The opposite charge was induced by the appearance of negative charge on ground aluminum foil connected to the high voltage sphere of a Van de Graaff generator. (a) a diagram of the experiment (b) a photograph of the same experiment [See also front color plates and Figure 1.6 (a)].

Attempts have been made to attribute the story in the *Ansei Chronicle* to electrostatic charges produced on the nails by charged aerosol, as was done for a report from a watch repairman in Italy, who claimed a tiny watch mechanism jumped into the air just before an earthquake (Tributsch, 1982). Because the hypothesis of charged aerosol could also explain these results, such an effect was eliminated by placing the magnet and nails in a plastic box to shield them from any ion wind. The iron nails still swung and dropped during nearby electric discharges from the Van de Graaff generator, meaning that electric induction was sufficient explanation for the behavior of the nails.

9.2.2 Disturbed orientation of a magnetic compass

Statements:

Small compass magnets have been said to have fluctuated violently before the Eastern Nankai and other earthquakes. The anomaly has been rather dubiously attributed to cosmic radiation, explained thus:

> *The earth's magnetic field changed before the earthquake so that cosmic radiation penetrated to the non-polar area causing earthquake light similar*

*to the auroral light caused by ionization by charged particles from solar
flares...*

It could be added that odd hypotheses like these appear because no serious attempt
is being made to bring science to bear on unusual events before earthquakes. Although the author might also be accused of producing odd hypotheses, his efforts
are at least a serious attempt at an objective scientific analysis of precursor reports.

Experiment:

A compass was placed near an antenna and a pulsed field applied. Naturally enough
the compass needle moved with the interaction between the magnetic needle and
the electric field. When systematically varying low frequency EM sine waves were
applied, the needle spun round.

Explanations:

As discussed, changes in the earth's magnetic field before earthquakes are negli-
gible: not enough to draw increased cosmic radiation into the earth's atmosphere,
and certainly not in amounts intense enough to cause unusual animal behavior before earthquakes. If that were the case animals would have skin burns, as in overex-
posure to X-rays or intense UV light.

A change of a few nano-Tesla (nT) in the Earth's magnetic field before big
earthquakes is about one ten-thousandth of the Earth's magnetic field. Given the
frictional forces impeding rotation of the compass needle, such a small change
would not be sufficient to move the needle. The fluctuation is much more explicable as an effect of charges induced by an electric field.

9.2.3 Car navigator: Fluctuating direction arrow

Statements:

Drivers claimed that their car navigators did not function—or malfunctioned—a
day before the Kobe Earthquake. In one case the navigational arrow swung 180
degrees from the true in a car in the city of Kobe. Generally navigators did not
function normally before and after the earthquake, but returned to normal a month
or so later.

Experiment and Explanation:

Modern car navigators are set up to receive satellite signals coded as 0 and 1 to cut
out background noise. Electric discharges were made using a Van de Graaff generator near such a modern navigator. Line noises caused by EM pulses were observed
on the screen, but there was no fluctuation in the arrow.

When the author spoke to a woman who provided such a report she said that the
navigator in the vehicle did not use the GPS system but an older system based on a

magnetic compass. Hence, EM waves at low frequency may have been responsible for the change in the arrow's direction on the screen of the navigator, due to an induced electric field.

9.2.4 Mischief by an invisible man?

Swinging iron chains:

Two hours before the Eastern Nankai Earthquake in 1944, iron chains hanging from the ceiling in a military factory began to swing, knocking a nearby electric furnace and making clanging sounds. [This was not coincidental with the vigorous movement of the bubble in the surveying level (Chapter 8).] The phenomenon was reproduced in the laboratory by inducing charge on small iron chains hanging near the high voltage sphere of a Van der Graaff generator. The chains swung towards the sphere.

So it is possible that an intense electric field, presumably electric pulses, may have appeared before the earthquake inducing charges on metal objects so they were attracted to each other, causing swinging in the iron chains.

Arc discharges between iron bars:

A day or two before the 1976 Tangshan Earthquake (M8.2), China, arc discharges were observed between iron bars lying on the ground—as if an invisible man was using an arc-welding machine (Dai, 1996). (A small arc discharge caused by static electricity in a human body may be seen sometimes when one unlocks the car or touches a doorknob.)

An intense electric field is formed between iron bars (electric conductors) if charges appear on the ground. Opposite charges may appear at the ends of the bars producing sparks by atmospheric breakdown at 3 MV/m.

A charge appearing on the ground intense enough to swing iron nails and iron chains and move magnetic compass needles would also explain arc discharges between the iron bars. The charge density of less than one millionth of a Coulomb per square meter on the ground would be sufficient to cause discharges at the sharp edge of iron bars. (This is less one thousandth of the charge necessary to run a small piece of consumer electronic equipment for one second.)

9.3 Unusual behavior of electric home appliances

9.3.1 Faint glow of fluorescent lamp

Statements:

Fluorescent lamps glowed faintly before the Kobe Earthquake (Wadatsumi, 1995). They also lit up spontaneously before the Tangshan Earthquake (Dai, 1996). The author has also seen a faint glow from a fluorescent lamp in a room during a thunderstorm.

Experiment:

A fluorescent lamp held close to the charged sphere of a Van de Graaff generator in the dark, lit up because of the intense electric field. A plastic sheet charged by frictional electricity also produced a glow from a fluorescent lamp (surprising pupils in a lecture on earthquake precursor phenomena). Air-gap discharges between the high voltage sphere and an electrically grounded rod generated pulsed EM waves that also lit up the lamp in a darkened room (Figure 9.2). A fluorescent lamp exposed to EM waves at 1 MHz from a Tesla coil lit up as it also does in a microwave oven. (A Tesla Coil is a high voltage transformer that generates very high voltages at high frequency.)

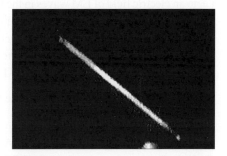

Explanation:

In a suitable environment, EM waves from thunderstorms (produced by atmospheric lightning) light a fluorescent lamp. The faint light of a fluorescent lamp before earthquakes could similarly be due to an electric field of pulsed EM waves.

Figure 9.2 A fluorescent lamp in a laboratory lit up when it was exposed to EM pulses, as fluorescent lamps have been reported to do during thunderstorm lightning and before earthquakes.

9.3.2 Radio interference

Incidents:

Truck drivers reported radio interference on a highway near the epicenter in Kobe about 5 a.m., 45 minutes before the Kobe Earthquake. Some were unable to get any reception at all. Before the Tangshan Earthquake factory workers manufacturing radios were unable to tune in any broadcast signals. The phenomenon seemed limited to that area, and reception returned to normal after the quake (Dai, 1996).

Experiments:

A transistor radio placed on the high voltage sphere of a Van de Graaff generator went dead at a high voltage but worked normally when the sphere was grounded. Air-gap discharges between the sphere and a grounded rod caused interference on another radio nearby (Ikeya and Matsumoto, 1998). FM broadcasts were not affected during these experiments. (A radio receiver is a good sensor of electromagnetic waves; a small radio placed near a computer will pick up EM interference from the computer.)

Explanation:

The air-gap discharges produced EM interference in the radio wave frequency range. Radio receivers tuned into AM bands are easily affected by EM noise; their longer, medium waves are more subject to interference than VHF FM waves. (Those seeking to use radio broadcasts for earthquake forecasting should obviously tune in to AM (rather than FM) broadcasts.)

Reported interference in FM broadcasts has been attributed to scattering of broadcast VHF waves according to a reasonable inference that the ionosphere descends locally from its usual height in response to the appearance of charges on the ground i.e. to an electric field of EM waves in the ULF range (Kushida and Kushida, 1998).

The reflection of FM waves might, rather, be caused by a change in the atmospheric conductivity around the epicenter due to charged aerosol. Whether the ionosphere is actually affected or not has yet to be established. For more on the apparent lithosphere-atmosphere-ionosphere (LAI) coupling, see Chapter 11.

9.3.3 "Barber-pole" color and speckle interference on TV screens

(a) Reproduction of barber-pole noise

Video films recording precursor noise: A video recording made 40 km away from the Kobe Earthquake epicenter of an NHK program broadcast at 11.18 p.m. (about six and half hours before the Kobe Earthquake), showed ghosting, "barber-pole" color and white speckle interference lasting for about 30 seconds. The broadcast and received images are shown respectively in Figure 9.3 (a) and (b). At this time hamsters were reported to be agitated and biting each other and mice were much more active than usual (Matsumoto *et al.*, 1998) and Figure 5.5 (a).

A person living in Kyoto, about 80 km away from the Kobe epicenter, supplied another video film that recorded the same section of the program. In this case no "barber-pole" noise was recorded but flashing lines appeared across the screen at about 11:18 p.m. No seismic waves were detected at the time, so low frequency EM pulses (not detectable by seismographs) generated by local fractures in stressed rock close to the observation site might have been the cause.

However, the times did not exactly correspond, so a time difference needs to be accounted for. A speculative explanation might be that radiowave interference (generated by pre-seismic EM waves) was emitted either heterogeneously (in a focussed beam)—maybe because of some inexplicable polarization from a ferroelectric orientation of electric dipoles (Ikeya *et al.*, 1998)—or the EM waves moved through the crust at an extraordinarily slow speed (Section 11.3.8, Table 11.2).

In any event, only the low frequency ULF component of EM pulses can be propagated from underground to the surface without attenuation (due to the large skin depth of these waves) and induce charges on the ground around fault planes. Appearance of this intense charge on the ground would cause electric discharges everywhere, generating EM noise in the high frequency (TV broadcast) range. They might also create disturbance in the ionosphere, creating high frequency EM waves by some unknown non-linear mechanism.

Analysis of videotapes:

Waveforms and signal spectra in both video recordings of the film were analyzed using a digital storage oscilloscope (DSO). An image on a TV screen is composed of 525 horizontal lines that are scanned from left to right in 0.063 ms by the NTSC system used in the USA and Japan. The luminosity signal (E_Y) and sub carrier color difference signals (E_I and E_Q) at a frequency of 3.58 MHz are related to components of the red, blue and green signals (E_R, E_B and E_G).

Figure 9.3 (c) and (d) respectively show the luminosity (brightness) and color difference signals isolated from the distorted image (b) recorded before the earth-

Figure 9.3 Screen distortion during a TV program broadcast by NHK at 11.18 p.m., January 16, 1995, about six and half hours before the Kobe Earthquake. (a) The image without distortion. (b) An image distorted by speckle noise and "barber-pole" interference. (c) Luminosity image E_Y isolated from (b) and showing speckle interference. (d) Image of color difference signal E_I, isolated from (b) and affected by "barber-pole" noise (Matsumoto *et al*, 1998). For a color view of "barber pole" noise, see the front color plates. (Videotape, Mr I. Kawakata.)

quake. Only speckle noises appear in (c); there is no effect on luminosity. Barber-pole noise appeared in the image showing color difference signals (d).

An attempt was made to reproduce barber-pole noise in a regular TV program broadcast by exposing the TV antenna to EM sine waves from a microwave synthesizer. A color shift appeared at 215 ± 5 MHz (within the TV broadcast frequency range).

It seems reasonable to conclude that EM pulses composed of EM waves at a wide range of frequencies before quakes can create interference on TV screens.

(b) Reproduction of speckle noise and channel fluctuation

There was a report of a faulty channel selection function in a TV remote for several days at the epicenter area before the Kobe Earthquake and we wondered if replication of channel-selection malfunction might be possible in a laboratory using EM waves. Speckle noise also seemed a likely candidate for an EM wave experiment given that the length of speckle interference in the video (based on the horizontal-scanning rate of 63 microseconds) was estimated to be of the order of microseconds, the same as pulsewidths observed during thunder, ten minutes before a local earthquake, and also during the granite compression experiment described in Chapter 5.

Experiments:

A TV antenna was exposed to EM waves produced by a microwave synthesizer and by air-gap discharge from a Van de Graaff generator. The discharges created speckle noise. A portable TV with a liquid crystal screen and a standard TV set were placed, respectively, on the sphere and also one meter from the sphere, during the air-gap discharge. Channel setting anomalies and noise were observed on the portable TV (Figure 9.4) on the sphere and in the standard set at a meter distant, during an arc discharge from the Van de Graaff generator.

Explanation:

Digital noise created by the EM waves scrambled the digital channel selection code in the TV and interfered with the broadcast frequency.

(c) Preseismic and coseismic EM pulses?

It is interesting to note that a small earthquake of M2.1 occurred at 11:49 p.m. on January 16 at the epicenter, apparently one of the four Kobe foreshocks. However, an episode of fluctuating channels at the epicenter area (above) occurred 30 minutes beforehand, meaning it can't be linked to the foreshock itself. In addition no such interference had ever been reported in connection with frequent small earthquakes in the area. So the incident is puzzling unless it was in response to presismic EM waves generated locally by increasing fault stress.

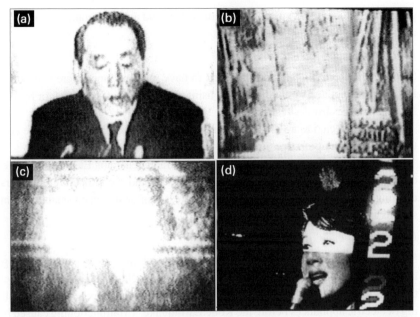

Figure 9.4 The effects of EM interference in an image on two color TV sets. (a) Normal image (b) Loss of image on the high voltage sphere of the Van de Graaff generator. (c) Horizontal line noise and "snow-like" speckle noise produced by arc discharges (d) Noise and an unstable channel setting (Channel 2).

There was one report of distorted TV images a few seconds before the Kobe Earthquake affecting not just one but at least two channels, followed by earth sounds and then the tremor. This apparently indicates coseismic generation of EM waves producing interference, and then earth sounds, just before the arrival of S-waves (the quake).

Japan has just implemented a hi-tech digital broadcasting TV system that gives a high-resolution color picture, but the new technology will still remain susceptible to color changes and other interference from EM noise generated by earthquakes or lightning. Even cable TV will be susceptible.

9.3.4 Strange sounds from refrigerators, spontaneous on-and-off switching of appliances

Statements:

Reports of mysterious buzzes and alarm sounds over intercoms, and strange sounds from refrigerators and air conditioners are among precursor reports collected after the Kobe Earthquake (Wadatsumi, 1995). Kobe citizens were also surprised and puzzled by spontaneous switching of radios, TVs and air conditioners around 2 - 3 a.m. before the earthquake at 5:45 a.m.

Experiments:

Some of the electric appliances reported to have malfunctioned were exposed to EM pulses generated by arc discharges using a Van de Graaff generator or by 100 MHz EM waves from a Tesla coil.

A TV connected to a video recorder, which could be operated by remote control, switched on upon exposure to EM pulses.

A refrigerator was charged using the Van de Graaff generator. Electric discharges occurred in the circuitry affecting internal devices and the temperature sensors, creating malfunctions that switched the compressor off and then on again, creating sounds and vibrations in the refrigerator.

Explanation:

Modern electrical equipment employs a simple switching system using digital signals. Mains power is still connected to this quick switching apparatus even when the switch is off, meaning preseismic EM pulses could cause spontaneous switching of electric appliances before the Kobe Earthquake. (Malfunctioning caused by EM waves will not occur if the appliance is not connected to the mains socket. Only a fluorescent lamp operates differently.)

Buzzing sounds from intercoms and cellular phones may also be caused by EM interference. In fact, spontaneous switching and malfunctioning of nearby electric appliances in our laboratory surprised us sometimes during our experiments using a Van de Graaff generator and a Tesla coil e.g. a laboratory oven at room temperature began to warn us that it was overheating and a TV 2 m away switched on.

Intermittent short circuits within the refrigerators would have turned compressors off and on causing the sounds and vibrations reported before earthquakes and reducing cooling efficiency. This could explain reports of spoiled yogurt; certainly the direct effect of EM pulses on lactate cells is doubtful since the refrigerator would electrically shield the yogurt.

The reported malfunctioning of radios, TVs, air-conditioners and intercoms at midnight before the earthquake is consistent with preseismic discharges and effects of intense pulsed electric fields such as earthquake lightning. The unscheduled broadcast of the Koranic prayer from a local mosque one hour before the Izmit Earthquake in Turkey may have been due to spontaneous switching on of an electronic device. It should be noted that, at about the same time, budgerigars were very agitated, and frightened children woke people up according to reports (Section 2.3.3). These are consistent with the appearance of intense EM pulses in the epicenter area before the Izmit Earthquake, just as before the Kobe Earthquake.

9.3.5 Activated mobile phones with no call records

The big surge in popularity of mobile phones occurred only three to four years ago so there are not many reports of malfunctioning mobile phones before the Kobe Earthquake in 1995. However, there were reports from both Kobe and Izmit of mobile phones whose call indicators had activated but had received no calls. In other words, the display lit up and there may or may not have been a ringing tone, but there was no record of any caller (Figure 9.5).

Experiment: EM noise created by electric discharge

Figure 9.5 Mobile phones exposed to electric discharges of EM waves lit up and rang but left no record of any caller.

Ten mobile phones manufactured by different companies were placed on the Van de Graaff generator and the grounded sphere was placed close to the high voltage sphere. Electric discharges generated EM pulses. In two of the ten phones the display lit up and the phones rang. (The others may have been better shielded.) Again, lay reports of mobile phones receiving calls but leaving no record of a caller may be explained as an effect of preseismic EM pulses.

9.3.6 Assorted unusual reports

(a) Burglar-proof alarm

A car security alarm, which used an electrostatic field, sounded at 9.30 p.m., January 16, about 8 hours before the Kobe Earthquake—about 1hr 45 mins before the TV picture interference captured on video. The incident may be another case of spontaneous switching on of an electronic device.

(b) Cat disappears from locked car

A cat mysteriously escaped from a locked car before a large aftershock in Turkey (Ulusoy and Ikeya, 2002) while its owners were asleep inside the car. Woken by the tremor they found the cat outside the locked car. It is possible that the automatic power window, activated by intermittent EM pulses, opened briefly, letting the cat out, and closed again.

(c) More computer crashes?

There was a statement that personal computers were more troublesome than usual just before the Taiwan-921 earthquake. If pulsed noise affects only one bit in the digital code, a calculation result can be completely skewed, but usually there is sufficient redundancy built into system software to absorb these fluctuations. The author didn't notice anything unusual about the performance of his computer before the Kobe earthquake, but then again it crashed so frequently—with a high voltage line 30 m away and a bus station closer still—it would have been hard to know! (Data on the frequency of crashes would have to be kept if computers are to be used in earthquake forecasting.)

(d) Odd sounds from a power transformer

Two electric power transformers made strange "hoh-hoh-hoh" sounds from 13 to 6 hours before an earthquake (M5) in Canton, China in 1970 (Rikitake, 1979). Such sounds are produced from a transformer by coil vibration when there are minor voltage and amp instabilities in the power line, the instability having been induced in the power lines from outside.

Dr T. Higuchi (General Research Institute of the Kansai Electric Power Company) has detected noise before earthquakes by measuring current in the neutral line of three-phase power lines. Power lines are effectively huge antennae, or circuits, capable of measuring earth potential differences, and are therefore possible detectors of earth current and/or EM waves before earthquakes.

9.4 Mysterious clocks and the "Alice in Wonderland" Syndrome

9.4.1 Did time stop or speed up?

There was an anecdotal report that clocks in a local stock exchange did not show the correct time before the Kobe earthquake. Quartz clocks in Beijing, at a distance of 160 km from the epicenter were reported to have stopped eight hours before the Tangshan earthquake (M8.2) in 1976. The same clocks also stopped before large aftershocks (Dai, 1996).

A newspaper reported that a medical doctor in Osaka had a radio-clock which adjusted its time every hour in response to a 40 kHz radiowave signal from a Government telecommunications laboratory, close to Tokyo. His clock lost two seconds before the Kobe earthquake, but ran to time again after the earthquake.

A citizen of Ashiya City, 25 km away from the Kobe epicenter, telephoned the author and said that the hands of her quartz wall-clock stopped moving and then moved slightly backward for a few seconds a day before the Kobe earthquake. They resumed normal speed and function after the earthquake. She wondered if she had been "seeing things" because nobody believed her, and she sought a scientific explanation. A girl student who also claims she saw the hands of a clock move rapidly before the Kobe earthquake requested confidentiality in case people thought she was a little strange.

The "Alice in Wonderland" Syndrome?

According to the theory of relativity, time may pass faster or slower. One cosmological theory says time may even reverse if the universe starts to contract. Obviously erratic clock malfunction does not fall into the category of a real time anomaly, but to someone who hears reports of hands spinning forwards, backwards and jerking to and fro, it can sound like a case of 'Alice in Wonderland' syndrome (Burstein, 1998; Mizuno *et al.*, 1998), in which symptoms of nausea and vision disturbance accompany a recurrent throbbing headache on one side of the head. Those with the syndrome have a disordered perception of time and space. The disorder can also be induced by drugs and gained its name from the book *Alice in Wonderland* by Lewis Carroll (1953).

In *Alice in Wonderland* (Chapter 7), the Hatter, who has become mad through exposure to toxic mercury used in tanning leathers to make hats, says to Alice at the Mad Teaparty,

> *Now if you only kept on good terms with him (Time), he'd do almost anything you liked with the clock. For instance, suppose it were nine o'clock in the morning, just time to begin lessons: you'd only have to whisper a hint to him (Time), and round goes the clock in a twinkling! Half-past one, time for dinner!*

There have been suggestions that people who claim to have seen clock hands jerk backwards and forwards or move at increased speed before earthquakes had some sort of psychological condition at the time. Did they?

Experiments:

A quartz clock was placed on the high voltage-sphere of a Van de Graaff generator (See Figure 9.6). No change in the movement of the hands was observed as the sphere was charged. However, the hands stopped moving when an air-gap discharge was made to a grounded rod close to the high voltage sphere. The alarm rang and then the second hand rotated at a speed eight times faster than normal when arc discharge occurred between the plastic case and the high voltage sphere.

Figure 9.6 The second hand of a quartz clock rotates at eight times normal speed on the high voltage metal sphere of the Van de Graaff generator.

An exposure to EM waves from an antenna at an emitting power of 10 W in a frequency range from 120 to 130 MHz stopped the clock. At the same time the automatic focusing motor of a video camera taking a picture of the clock at a distance of 2 m malfunctioned, blurring the image. The intensity of the EM waves was about 2 W/m^2 (30 V/m) assuming uniform radiation, a little higher than that causing unusual animal behavior in budgerigar and mice, as described in Chapter 4.

Explanation: Short circuits in a digital integrated circuit

The hands of a quartz clock are driven by a stepping motor, itself driven by electric pulses fed from a digital integrated circuit (IC) composed of flip-flop circuits in a state of either "0" or "1". The clock's plastic case was charged positively by collecting charges from the air and the clock's insulation was broken down by air-gap discharges. This discharge state generated electronic noise, which disturbed the digital circuit (essentially creating a short circuit by interfering with the digital code, and producing errors, thus affecting the generation of pulses, motor and hands).

Sometimes, discharges of accumulated charge in a printed circuit will also result in short-circuiting. Such short-circuiting would cause spontaneous ringing of a buzzer.

The eightfold increase in the speed of the clock in the present experiment suggests that three flip-flop circuits in the digital IC were short-circuited giving eight times the number of electric pulses to the pulse motor, resulting in the faster movement of the second hand (the loss of each flip-flop doubling the speed of the second hand).

Thus, the statements about rapidly moving clock hands before the Kobe earthquake appear to have nothing to do with the "Alice in Wonderland" Syndrome (Ikeya *et al.*, 1998a), but were probably real.

It should also be repeated that preseismic EM waves are not 120-130 MHz or any particular frequency but are pulses composed of many different frequencies, so having a much higher probability of affecting circuits than if they were one specific frequency.

So, the reports of aberrant clocks before the Kobe, Izmit and Taiwan earthquakes could have been real natural phenomena.

9.4.2 Dead phosphors on a wristwatch

A man passing through a tunnel near Izu (close to Tokyo) in a train looked at his wristwatch. He was unable to read the phosphorescent markers and concluded the phosphors had failed. This happened on August 29, 1923 and the Great Kanto earthquake occurred two days later on September 1. On September 2 the man found that the phosphorescence had reappeared (Kamei, 1976).

Experiment:

When a light emitting phosphor was placed under an intense electric field generated by the Van de Graaff generator, in a dark room, the phosphorescence faded and then disappeared. It took hours for it to be restored.

Explanation:

In the 1920s phosphors in wristwatches were excited by alpha rays emitted by radium. Alpha rays have a strongly ionizing effect, creating electrons and holes (Figure 7.4). The slow recombination of electrons and holes in the phosphor was responsible for the phosphorescence. An electric field removes ionized electrons so reducing the efficiency of the electron-hole recombination in the phosphor. Hence, phosphorescence may be degraded under an intense electric field before an earthquake. When the field disappeared the distribution of the electrons would return to normal in time.

9.5 Earthquake forecasting using electric appliances?

By exposing radios, TVs and air conditioners to EM noise at about 10 W/m² or 60 V/m in many laboratory experiments we have reproduced the spontaneous switching of these devices and appliances that have been reported before earthquakes. Essentially, the phenomenon of malfunctioning modern electric home appliances before earthquakes (Figure 9.7) may be explained largely in terms of short circuits

Figure 9.7 Electric home appliances—present and future—that can be expected to malfunction on exposure to EM waves.

from digital data errors produced by electronic interference from intense EM pulses, though some exaggeration or invention of reports cannot be excluded.

However it is risky to use electronic devices in earthquake forecasting. There are too many other factors producing interference to make them reliable indicators of an imminent large quake e.g. a sporadic E layer in the ionosphere formed by solar flares (and maybe also by the apparent bending of the ionosphere towards intense ground charges in epicenter areas), globally disturbs the transmission of EM waves creating interference in radio and TV reception.

Radio and TV transmitters and receivers are not always maintained well, nor is the power supply universally stable, and modern environments are noisy with sudden EM discharges, so malfunctioning electric appliances are a fact of life for many people. PC crashes are too frequent and commonplace to be used in prediction. Consumer demand for better-performing appliances might eventually make them more reliable indicators of imminent earthquakes, but electrical shielding of circuitry will no doubt also improve.

In any event, if greater efforts are not made scientifically to predict earthquakes, lay people may be able to get enough warning from sudden malfunctions in a group of ordinarily well-performing electronic appliances to make sure life-saving and property-protecting precautions are in place in case of a quake.

9.6 Care needed to avoid interference in data measurements

Many university geophysicists have tended to do their field work with cheap recorders and electrometers that are not made for regular outdoor use i.e. the manufacturer's specifications stipulate use in air-conditioned rooms only. This introduces a risk of electronic noise and errors. Perhaps the best-known example of interference was the illusory discovery of "cold fusion." In the late 1980s, scientists thought they had found a way to produce energy by nuclear fusion at very low cost by putting a current through heavy water at room temperature using special palladium and platinum electrodes. But they were apparently misled by electronic interference created partly by moisture precipitation on the circuit from the bubbling humidifying action of the heavy-water cells. When they gained grants because of their "discovery" and moved the operation into dehumidified rooms their cold fusion discovery burst like the bubbles. They were unable to reproduce their result.

These cases are warnings to researchers, especially non-professionals, working in short term forecasting that the greatest care needs to be taken in measuring seismo-electromagnetic signals (SEMS) because their equipment could be subject to electronic interference—not least from EM waves themselves—and misleading associations made.

Necessary caution is one side of the story in EM/earthquake research. The other side is an unwillingness or inability to see past an entrenched point of view. When the laboratory response of a quartz clock to EM waves was described at a conference of the International Geophysical Union (IGU) at Birmingham, England in 1999, a leading geophysicist simply said "Rubbish! If a quartz clock in my more than three hundred seismographs in California were to have malfunctioned, the earthquake focus could not have been determined. We have not had an event like that in the last twenty years!" The man dismissed the evidence outright because he leapt to the defence of the clocks in his system (presumably well-shielded expensive quartz clocks used in precise scientific instruments) and failed to see that cheap quartz clocks in plastic casing fell into a different category altogether. In so doing he failed to hear a legitimate case for EM effects on electronic objects.

In another, but much smaller, symposium a leading geophysicist said he had considered certain seismograph clock data to be electronic error and omitted the data from his results because no other local stations gave similar data. The author upset him by suggesting the omitted data might have been a meaningful EM pulse signal before the Northridge Earthquake and that the settings of their instruments (viz. their low sampling frequency and the time window of their A-D converter) meant the probability of two instruments picking up the same pulse would have been very small. A skeptical referee also argued that electronic error might have been the cause of the high activity levels reported in mice before the Kobe Earthquake (Chapter 5).

Strange data should not necessarily be blamed on electronic error; they could be caused by preseismic EM interference.

9.7 Summary

Malfunctions in a range of modern electronic appliances, reported before four large earthquakes in 1995 and 1999, have been reproduced in laboratory settings by exposing them to EM waves, giving support to the emission of EM waves both before and at the time of earthquakes.

Though EM interference can be produced from many sources, lay people should be alert if they notice in a number of domestic electric appliances, over a short period of time, odd functions of the kinds that have been described in this chapter. An earthquake may follow, it may not, but it would be a good time to make sure that precautions have been taken to save life and property in earthquake-prone areas.

10

Forecasting using Animal Monitoring

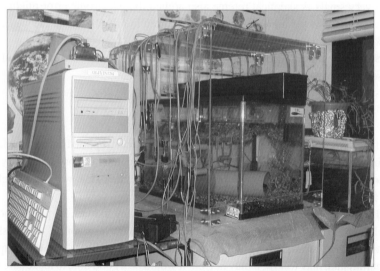

Simultaneous observation of seismo-electromagnetic signals (SEMS) and unusual animal behavior. Until reliable SEMS detection systems are in place, and
even when they are, automated observation of unusual animal behavior will be a
useful adjunct to other detection systems—just as sniffer dogs are in drug
detection.

10.1 Introduction

Numerous laboratory experiments have shown that plants, animals and even electronic objects respond to EM waves. In a significant number of cases the behavior corresponds to that noted in legends and reported in four >M7.3 earthquakes in the last decade.

But it is something of a leap from the laboratory response of animals to EM waves to earthquake forecasting using animals. This chapter reviews the use of two animals in earthquake forecasting. Obviously two is not many, but the field is very new.

Particularly the chapter looks at the catfish—partly because it has a long and legendary association with earthquakes, partly because it has a proven, clear response to EM waves (and its response behavior can be unambiguously differentiated from its natural motionless state, unlike the more sensitive eel's), and partly because a catfish and EM monitoring program is already in place in some Japanese schools, providing a prototype of how such a monitoring system might be set up.

It is preferable that animal monitoring and EM monitoring should be carried out simultaneously (in other words that e.g. the catfish aquarium and EM detection apparatus function as a unit), so that animal behavior can be numerically checked against recorded EM signals. The catfish monitoring program is set up in such a way.

Although mice have not been used in forecasting, the chapter suggests their potential might be significant given evidence of their much higher activity rates for several days before the Kobe Earthquake, in conjunction with other EM anomalies measured around that time. The chapter also looks at attempts in China to use the parakeet.

As this book has attempted to demonstrate, unusual animal behavior before earthquakes can be an EM anomaly. If sufficiently extensive simultaneous monitoring networks are set up, legendary and reported links between earthquakes and animal behavior can be further evaluated.

10.2 Parakeets at a Chinese high school

A teacher at one of China's Railroad High Schools, Mr T. Zhang, has been forecasting earthquakes using parakeets, microphones and audio frequency filters for more than 20 years. His apparatus records high frequency twittering sounds and low-frequency wing fluttering, thus detecting unusual disturbance in the birds' usual patterns. The number of sounds beyond a certain threshold is recorded in 24-hour periods.

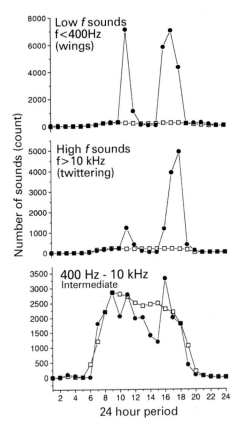

Parakeets usually make more sounds in the morning and in the evening. However, on occasions, as shown in Figure 10.1, there is an unusual increase in twittering and movement sometimes at night. Eight days after the activity recorded in Figure 10.1 an M6 earthquake occurred, at an epicenter distance of about 100 km.

Professor Li at Beijing Technical University observed an increase in parakeet activity eight days before the Kobe Earthquake (M7.3) in 1995. The eight day period is intriguing, and has a general correspondence with some other precursor phenomena e.g. Loma Prieta, and catfish movement (See Section 10.4.3), but may be purely coincidental. We could speculate that the birds may have sensed EM precursor pulses associated with the Kobe Earthquake in Beijing, but this is unlikely: Beijing is 2000 km from Kobe. Or the birds may have simply responded to thunderstorms prevalent in the Sea of Japan at that time, or even, possibly, to accumulation of stress in rock, creating microfractures and EM emissions somewhere near Beijing.

Figure 10.1 Daily variation in low-frequency sounds (fluttering), high-frequency sounds (twittering) and intermediate frequency sounds in a cage of parakeets in China. Black squares indicate hourly averages and black dots show an anomalous increase 7-8 days before an M6 earthquake 100 km distant. White squares give background averages (Supplied by Mr T. Zhang).

10.3 The potential predictive value of mice

Mice are one of the world's most investigated mammals, routinely observed in laboratories all over the world. If data of their activity levels in earthquake-prone countries were made accessible to those working in the SEMS field, mice may be shown to have predictive value when considered alongside other EM data. Figure 10.2 shows the mice activity described in Section 5.3 and Figure 5.5, plotted along-

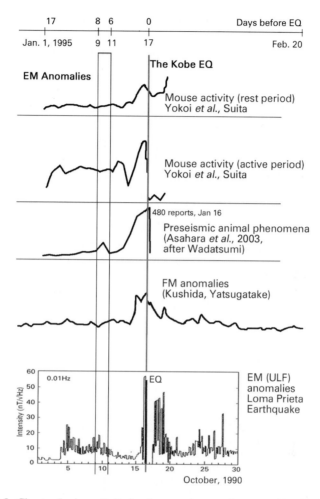

Figure 10.2 In Chapter 5 mice activity levels were shown to have considerably increased several days before the Kobe Earthquake. In this figure the levels are time aligned with reports of animal behavior and FM anomalies observed before the quake. The Loma Prieta EM anomalies at the bottom of the figure show the same trend.

side other EM anomalies recorded before the Kobe Earthquake. The figure should also be viewed alongside Figure 11.8 (further EM anomalies). When Asahara's data in Figure 10.2 are compared with ULF data for the Loma Prieta Earthquake, one could argue for a broad peak 7-8 days before both M7 earthquakes. All data show a sharp increase a few days before the mainshock and a clear EM (ULF) effect on animal behavior.

10.4 Can catfish be used in earthquake forecasting?

10.4.1 Observation for 16 years: Alleged success rate of 30 %?

The Tokyo Metropolitan Fisheries Experimental Station studied the correlation between catfish movements and seismic activities for 16 years from 1976. The catfish were violently active from eight to four days before an earthquake of M>5.9 (Figure 10.3), epicenter 86 km distant. Other unusual behavior was apparently observed from about ten days before earthquakes of a magnitude of more than M3 in 27 cases out of 87 (30 %) over a period of 13 years (Asano, 1998). Though the station made no attempt to use this material predictively, or to evaluate the "success" of correlation between catfish movement and earthquakes, the data came into the possession of people who sought to do so. Somehow a success rate of 30% was derived from the data, even though the theoretical relationship between magnitude and distance from the epicenter had not been applied to determine whether or not the forecasts were successful.

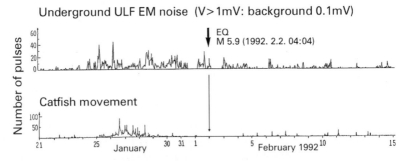

Figure 10.3 A correlation between ULF noise and catfish movement 8 days before a M 5.9 earthquake. Measured at the Tokyo Metropolitan Aquarium (After Fujinawa and Takahashi, Report of Earthquake Prediction, No. 52, 145-149, 1994).

During an annual conference of the Seismological Society of Japan the author was asked by a TV news reporter for a comment on the "catfish network" (Section 10.5) and the alleged 30% success rate of catfish. After an hour of filming, and seeking to avoid controversy, he remarked that, if the 30% figure were correct, catfish apparently had a better record than seismologists, who tended to make their comments after the event rather than before it. Unfortunately that was the only comment that made the TV news; the author's talk on practical preparedness for earthquakes was ignored, and not surprisingly, a few seismologists were not only stunned but became unfriendly as a result. Careless media comment can be damaging.

10.4.2 Correlation between EM pulses (ULF) and catfish behavior

The relationship between earth potential measurements and catfish behavior can be empirically established—as already shown in Figure 10.3. There appears to be some correlation between the movements and ULF emissions before the M5.9 quake even though the two detection sites were about 100 km apart (Fujinawa and Takahashi, 1994).

Any movement in catfish that coincides with an observed electric field effect can be explained as an electric field effect. Bear in mind that the normal behavior of catfish is to be motionless inside a pipe or other enclosed space.

10.4.3 Aquarium with digital storage oscilloscope (DSO)

To investigate whether aquatic animals respond in unusual ways to electric pulses, this laboratory linked a simple aquarium to a digital storage oscilloscope (DSO) to measure electric pulses in water.

Parallel plate electrodes were immersed in the aquarium with the water as a dielectric (Figure 10.4, left). To save computer memory the oscilloscope trigger level was set to 10 mV (10 mV/25 cm; 40 mV/m) so that the oscilloscope recorded only responses to a pulsed voltage in the frequency range 10 Hz to 10 MHz. Data were recorded at 40 million samplings/s and transferred to computer hard drive with a total processing time of 12 seconds. Data from pulsed electric fields before earthquakes and from lightning were included.

Before any animal was placed in the aquarium a number of different electric pulses were recorded. Figure 10.4 (a) shows the last of a series of five unidentified pulses occurring over about five hours; it occurred 15 minutes before an earthquake at 17.37 h, May 29, 1996 (M3.8, focal depth of 20 km, epicenter 15 km away). No coseismic pulse was observed. Atmospheric lightning causing electric pulses in the aquarium water, is shown in Figure 10.4 (b), and (c) shows artificial electric noise induced by switching on a ceiling fluorescent lamp in the room.

Preseismic pulses startle an eel and hamsters

Electric pulses of unknown origin are shown in (d). They were detected at the same time as eels and hamsters placed briefly alongside the apparatus were observed, quite by chance, to make sudden behavioral responses. The eels made startled movements and the hamsters began fighting and grooming.

Our experiments have shown that eels will move suddenly and hamsters begin grooming in an electric field of 0.2 V/m produced by electric discharges using a Van de Graaff generator, so the pulses were probably more intense than that. Minnows and goldfish, which respond to an electric field of more than 10 V/m (Ikeya *et al., 1996a),* did not behave in any unusual way, indicating that the field

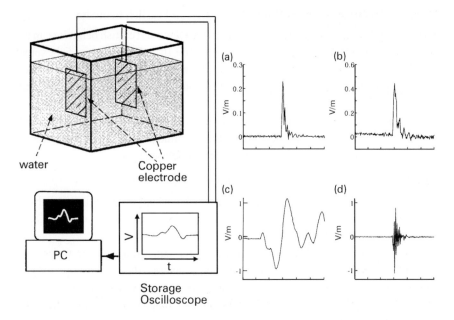

Figure 10.4 (Left) An aquarium with parallel plate electrodes and digital storage oscillo-scope (DSO). (Right) (a) An unidentified pulse that occurred 15 minutes before a local earthquake of M3.8. (b) An electric pulse recorded at the time of nearby atmospheric light-ning. (c) Electric pulses from a fluorescent lamp near the aquarium generated similar pulses in water. (d) Unidentified pulses which startled eels and hamsters; an earthquake (M3.9) followed 3.2 days later, epicenter 35 km distant. In all cases recording began after the trigger, in units of 0.0001 ms.

was probably less than 10 V/m. An earthquake (M3.9) occurred three days later, epicenter 35 km away.

Because of the eel's greater sensitivity we first used eels to investigate responses to EM waves, but they tended to leap out of the aquarium in response to pulses, and had to be retrieved from the floor. After six months we switched to the slightly less sensitive, and normally motionless, catfish.

10.4.4 Simultaneous observation of catfish behavior and SEMS: A case for successful forecast of the Western Tottori Earthquake (M7.3, October 6, 2000)

EM pulses and catfish movement were simultaneously monitored in this laboratory to investigate whether catfish respond to EM pulses before earthquakes. Electrodes were attached to a polyvinyl chloride pipe in an aquarium as shown in Figure 10.5 (a). A trigger level of 10 mV/m did not detect EM pulses produced by nearby light-

Figure 10.5 (a) An apparatus to monitor the bioelectricity of catfish. (b) Simultaneous observation of unidentified VHF pulses and monitored catfish activities eight days before the Western Tottori Earthquake (M7.2) on October 3, 2000. There is a clear correspondence at both 3:57:54 and 5:08:57.

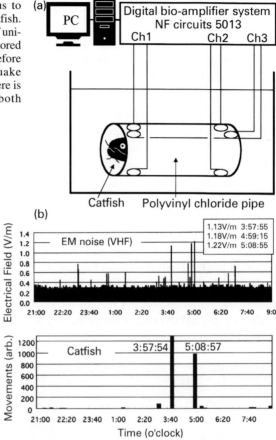

ning, so interference was minimized from this source.

The catfish, although remaining motionless in its pipe, moved its whiskers at the same time the monitors detected EM pulses. Such whisker movement was also observed when electric pulses with an intensity of 1 - 2 V/m and a pulse width larger than 0.1 ms were applied to the catfish (See Chapter 4).

The catfish moved violently in response to unidentified EM pulses eight days before the Western Tottori Earthquake (M7.2), at an epicenter distance of 130 km from the observation site. Figure 10.5 (b) shows a one to one correspondence between the movements and simultaneously observed EM pulses.

It was clear that the catfish was responding to electric pulses and possible that the unidentified EM pulses to which it responded so violently could be linked to an earthquake. On this basis the author predicted an earthquake somewhere in the area about eight days after the peak of the violent movement, which also coincided with anomalies in the lower to medium frequency wavebands that were being simultaneously monitored. Eight days later the Western Tottori Earthquake occurred and the data were added, at galley proof stage, to the Japanese book, *Precursors of Large Earthquakes* (Ikeya, 2000).

10.4.5 Catfish in turmoil before the Geiyo Earthquake (M 6.7, March 24, 2001)

Catfish were active in the Osaka Aquarium before the Geiyo Earthquake near Hiroshima, 240 km away from Osaka. Two other catfish were also observed in violent motion and a goldfish was seen leaping out of the water in the laboratory at Osaka University. Unfortunately no EM signal was detected at Osaka, although EM pulses were detected at a nearby observation site at Matsue, 150 km from the epicenter (Yamanaka *et al.*, 2002). After the successful forecast of the Western Tottori Earthquake the author predicted another about eight days after the peak of the active behavior and warned disaster prevention personnel and a nuclear facility. However, there was no earthquake eight days afterwards and apologies were in order. The unusually high level of activity lasted another two weeks and we finally concluded it was possibly seasonal, linked to unstable spring weather. Video recording ended.

However, violent movement and shock responses began again in the catfish about 20 hours before the Geiyo Earthquake (M6.7), epicenter 240 km distant, focus depth, 30 km (Figure 10.6). The video camera was turned on again, and at the time of the earthquake clearly records the trembling of the water surface and the catfish's pipe, and the excitement in the author's voice. The catfish, however, remained unperturbed.

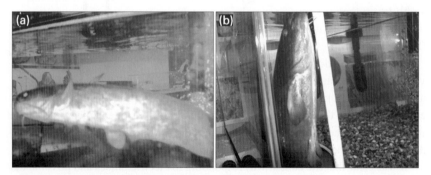

Figure 10.6 Photographs of violent movement in a catfish at 5:18 p.m., 20 hours before the Geiyo Earthquake (M6.7) at an epicenter distance of 240 km.

According to Mr Y. Kushida (Yatsugatake Astronomical Observatory), the Geiyo Earthquake was a plate boundary earthquake and the Western Tottori Earthquake an inland earthquake, caused by movement in a geological fault. Kushida, who claimed to have successfully forecast the Geiyo Earthquake based on his observation of broadcast FM radio waves, remarked that EM intensity typically peaks much more than eight days before plate boundary earthquakes. If this is so, it may be impossible using catfish activity to forecast plate boundary earthquakes well ahead; instead any turmoil might indicate a large quake is only hours away rather than days.

10.5 Catfish monitoring network: integrated science education

A 24-hour program is active now in some Japanese laboratories and high schools to monitor catfish behavior over much of Japan by computer and video camera (Asahara *et al.*, 2002). The image and range of activities are connected via the Internet to our laboratory in Osaka University (See Figure 10.7) and can be accessed by anyone at any time through the Internet using a mobile phone in image mode (http:// catfish.ess.sci.osaka-u.ac.jp/). The objective of this "integrated science education" project in high schools—essentially a "catfish network"—is to study:

- Social history (legends and folklore about "earthquake catfish" and other precursors)
- Behavioral biology (observation of daily, monthly, seasonal and annual rhythms)
- Electromagnetism (observing EM pulses and catfish behavior simultaneously)
- Earth Science (Seismology), (by studying earthquakes and disaster)
- Statistics (to correlate catfish activity with seismic activity or other e.g. seasons, magnetic storms)
- Information processing (by exchanging information over the Internet)
- Coordinated acquisition and analysis of data in real time

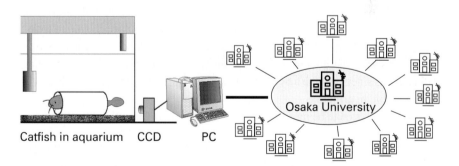

Catfish in aquarium   CCD   PC   Osaka University

Figure 10.7 An integrated science education program running in some high schools in Japan to simultaneously monitor EM emissions and catfish activities. The schools are connected to Osaka University and the Internet. Students learn scientific skills and how to sensibly interpret catfish behavior. At the same time the network gathers useful data on SEMS and animal behavior for specialist analysis.

Naturally catfish activity increases in spring following the south to north movement of the "cherry blossom front". Students in the program learn to determine whether catfish are responding to a spring rise in temperatures, the increasing length of daylight, to thunderstorms (lightning), the rainy season, or impending typhoons.

There are more than one hundred >M4 earthquakes per year in Japan. Before an M4 earthquake EM emissions could be expected to produce violent movement in catfish within a radius of 30 km from the epicenter. Obviously, the greater the number of stations (high schools) in the network the higher the chances of detecting violent catfish movement before earthquakes and, therefore, of successfully forecasting a large earthquake. The epicenter and magnitude of a forthcoming earthquake can be estimated by correlating the varying intensities of catfish behavior and electric fields in the observation area according to a theoretical equation (Ikeya *et al.*, 2000).

Information Technology may yet show that there is something to the old catfish legends.

10.6 From Parkfield cockroaches to animals fitted with meters

Dr C. de Groot-Hedlin passed a modern anecdote on to the author. Near Parkfield, California, an active fault has ruptured at fairly regular intervals—about every 30 years for the past 150 years. Many scientific instruments have been placed in the fault but it was not until one of the locals suggested animal studies should also be done that scientists agreed to collaborate in an animal experiment, using cockroaches. The resident was to feed the cockroaches during the course of the experiment, and the scientists to develop a computer program to monitor the cockroaches' activity. At some point in the exercise the computer indicated that all cockroach activity had ceased, and when the box was checked all the cockroaches were found to be dead of starvation. Nor were any earthquakes recorded during the period of the experiment. The attempt to involve animals was an abortive exercise, and Parkfield has since become a byword for those who believe earthquake forecasts to be impossible in principle.

But people need to be motivated to observe their animals, either by affection or to protect an investment. Cockroaches hardly fall into that category, but pets and valuable flocks and herds do. In this case there are already devices, which, fitted to animals for veterinary reasons, will not only help protect large investments in flocks and herds, but also provide valuable data that could be used to determine links between animal behavior and earthquakes. Such devices, similar to pedometers or cardiometers fitted to humans for health reasons, used by pet owners and farmers, would, if data were collected from a large enough group of animals, provide a general physiological average. Any data showing an unusual departure from these norms would be transmitted to the mobile phones of owners and specialists. Specialists would judge what large scale anomalous data might mean e.g. climate changes, the beginnings of an epidemic, or a possible quake. For instance, if the veterinary health of cows on Californian farms were monitored using embedded tips and transmitters, unusual behavior before earthquakes could also be detected. Dog pedometers are already available in Japan.

10.7 Forecasting using animals?

The author is of the view that animals should not be used, *per se*, for official earthquake forecasting. Rather, animal behavior should be objectively observed, along with observation of preseismic EM pulses and their source determination; analyzed numerically by specialists using dedicated computers and purpose-built instruments, and made available on an unofficial basis. Results are also of little use unless they are measured against distance from the epicenter and the time of precursor observation and magnitude. Obviously, changes in behavior must exceed normal behavior variability for the same year or for the same month in other years.

It is possible, in view of the amount of background EM noise in human environments, that EM data could even be filtered through known animal recognition of EM patterns. Certainly, until more reliable forecasts are possible, it is best to avoid dependence on any one method, and for disaster prevention on an unofficial basis automated computer monitoring of animals has a role to play.

Wadatsumi (1995) set up a Website PISCO (Precursory Quake Information System by Citizens' Observation) to collect information on possible precursors that could be used for earthquake forecasting. If reports increase in any area, a warning is posted on the PISCO home page. Obviously phenomena will occur from time to time that have nothing to do with earthquakes but the PISCO site is a useful reference for the public and repository for lay persons' observations of unusual phenomena. It also provides a resource for further scientific evaluation of reported phenomena.

But such initiatives are really more an experiment in people behavior than anything else e.g. are owners disciplined enough to check their animals at set times or informed enough to make objective assessments? They have to be highly motivated or else regimented. The Chinese claim to have successfully predicted the 1976 Haicheng quake by analyzing laymen's observations of animal behavior, but such a large-scale project requiring observational disciplines was probably only possible under the communist regime in place at the time. Today, computers have replaced the mobilization of citizens and in today's democracies a similar result can probably only be achieved by enlisting the help of computer-equipped schools.

10.8   Summary

It is clear that animals respond to EM waves and that EM anomalies in a range of frequencies have been recorded before earthquakes. It appears that some animals—catfish in particular, but also eels, hamsters and possibly mice, behave in unusual ways at the same time unidentified EM emissions are being electronically detected. This at least makes them candidates for any forecasting exercise, along with a number of other contenders, although obviously the work is in its embryonic stages. One of the best ways to establish whether animals—some more than others—can be seriously used in forecasting, is to continue to monitor animal behavior simultaneously with EM detection.

The catfish monitoring network in place in Japanese laboratories and high schools is an attempt to put such a system in place in a way that is educationally useful to students but also constructive from a research point of view. It is not impossible the network will produce enough data to allow the successful forecast of a large earthquake, as happened in the case of the Western Tottori Earthquake in 2000.

Although an argument can be made for SEMS detection using man-made materials with stable properties, rather than animals whose data cannot be absolute, pets and farm animals are ubiquitous and relatively easily observed. If behaviors are electronically monitored and professionally evaluated from large enough samples, the data may provide useful information for use on an unofficial basis. The technological age may yet prove that some of the old legends are not merely myths.

Reports of animal behavior to Internet Chat groups from interested citizens have their place as a resource to browse to view possible trends, but cannot match continuous electronic observation and specialist analysis.

Monitoring Seismo-Electromagnetic Signals (SEMS) - a General Survey

A composite photograph of SEMS detection equipment. Researchers in an increasing number of disciplines claim to have observed anomalous EM signals before earthquakes in a wide range of frequencies from ULF to VHF. Though the picture is confused in terms of mechanisms and it is very difficult to pin observation of SEMS down to specific earthquakes, the field is gaining increasing credibility.

11.1 Introduction

In the last two decades research into the relationship between earthquakes and changes in electric and magnetic fields has taken great strides and drawn on a broad range of disciplines: seismology, geodesy, geomagnetism, ionospheric and space science and telecommunication. The results are intriguing and promising, if not yet definitive. What is clear is that changes in electric and magnetic fields have been measured before earthquakes, and electromagnetic anomalies have also been clearly observed across a wide range of frequencies before earthquakes. What is not so clear is precisely how (and sometimes, if) they link with earthquakes and whether the complexity can be reduced to understood mechanisms with predictive value.

As a relative newcomer to SEMS the author does not presume to review the entire field. This chapter seeks merely to give a brief overview of the current state of research, the problems, and the possibilities. For scientists wishing to undertake further research into SEMS, there is an extensive literature in the bibliography.

11.2 Comments on changes in electric and magnetic fields

11.2.1 The proton magnetometer

(a) A proton magnetometer cannot detect EM pulses

The proton precession magnetometer is a device that measures the earth's magnetic field very precisely using nuclear magnetic resonance. Its data are considered to be extremely accurate. But when geophysicists first began to use this high precision instrument in the sixties they found it was giving much lower intensities for magnetic anomalies than had been estimated in the past from early measurements and legends. In other words the intensity of magnetic anomalies before earthquakes had been greatly overestimated before the introduction of the magnetometer (Rikitake, 1976; Park *et al.*, 1993) and in fact there was only a minuscule change in the earth magnetic field around the time of an earthquake (See Figure 11.1 (a), in which further data up to 1995 have been added.)

Understandably scientists then rejected the old data as being erroneous or misleading but in their adoption of the proton magnetometer they overlooked a significant shortcoming i.e. the proton magnetometer cannot pick up magnetic pulses faster than 1 ms and most electromagnetic pulses are faster. That is, in the Earth's magnetic field of 0.04 mT the nuclear magnetic moment of a proton precesses at 1 kHz. Any change in the field is estimated from the shift in the precession frequency indicated by the proton's nuclear magnetic resonance. The frequency of 1 kHz dic-

tates 1 ms for one cycle precession movement. So, un-
less the field persists for
longer than a millisecond, a
proton precession magne-
tometer cannot give an accurate field even in principle,
nor can it detect any pulses
of magnetic waves at fre-
quencies higher than 1 kHz
produced before earthquakes.

However the old-fash-
ioned analog apparatus comprising a galvanometric magnetometer and a recorder can
still respond to a pulsed magnetic field to some extent.

It is of interest to note that
the observed magnetic
anomalies accompanying
volcanic eruptions also decreased, but not as markedly,
after the introduction of the
modern magnetometer, as in
Figure 11.1 (b). This would
be simply because the variation of the volcanic magnetic
field is rather more static than
pulsed and the pulse width is
larger than 1 ms.

(b) Sampling frequency that
misses changes in pulsed
fields

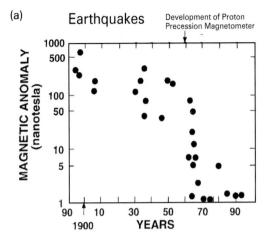

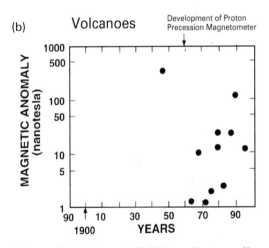

Figure 11.1 Magnetic field intensity anomalies before earthquakes and volcanic eruptions estimated from legends and measurements over time. There is a drop in intensities after the introduction of the magnetometer (Johnston, 1997).

It is a fundamental principle of signal detection that one must measure extremely rapidly to detect a fast pulse. The irony is that the sheer amount of data (millions of pieces per second) that needed to be stored if a full record of fast pulse measurements were to be kept overwhelmed early digital capacity, with the result that the digital process became slower than the old analog process had been. At some point a decision was made (that became a general modus operandi) to sample less often to conserve disk capacity, with the result that many fast EM pulses before earthquakes were missed. This goes some way towards explaining the lack of records of significant SEMS measurements—as conventionally observed—before earthquakes. However, these fast pulses can be detected with specialized equipment by raising the preset 'triggering level' to detect only EM waves above the trigger intensity. Recent developments in digital technology have revolutionized the quality of observational data, and new innovations will no doubt continue to do so. It is often the quality of the observational data (rather than the IQ of scientists) that allows the development of a new theory. This may help to remove scepticism about SEMS.

11.2.2 Atmospheric electric field

Changes in the atmospheric electric field were noticed as early as 1799 by Alexander von Humboldt in the Venezuelan Earthquake (Tributsch, 1982), (Section 7.5.3) and also during the Matsushiro earthquake swarm (Kondo, 1968). It appears that the nearer the seismic swarm activities are to the observation site, the clearer the decrease in the intensity of the atmospheric electric field. The decrease in electric field intensity before the Kobe Earthquake shows up clearly in Figure 11.2. The data were collected during testing of an electrostatic field meter for an instrument manufacturer, Kobe Radio Waves. Other events, like reported malfunctioning of electric appliances, TV and radio noise (Chapter 9), are added to the figure. So changes in an intense atmospheric electric field appeared to occur in Kobe before the earthquake. Although there was no lightning in the Kobe area at the time, there was lightning far away over the Sea of Japan to which the EM noise was ascribed (Izutsu (2003). But this is still puzzling.

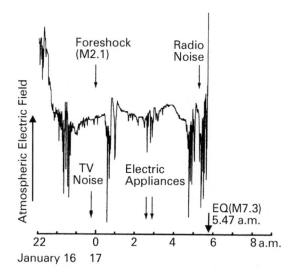

Figure 11.2 A record of the atmospheric electric field before the Kobe Earthquake on Jan. 17, 1995 (Data from Kobe RadioWaves). Malfunctions in electronic devices and appliances are entered in the figure at the times they were reported to have occurred.

Atmospheric electric field sensors, such as the Coronarm, manufactured by Central Lightning Protection, are being installed to detect electric fields (Kamogawa *et al* (2000).

11.2.3 DC geoelectric potential

Telluric current (TC):

Telluric currents are natural electric currents arising from charges moving to attain equilibrium between regions of differing electric potentials on and beneath the surface of the Earth. These earth potential differences have long been studied in earthquake prediction (Honkura, 1981) and can be recorded by observing differences in earth voltages between electrodes. Traditionally a voltage is measured and recorded between two electrodes, each 3 cm in outer diameter, 40 cm in length, buried 2 m deep and separated by anything between 10 m and 10,000 m. These are called short and long dipoles depending on the amount of separation between the electrodes. An electric field of 10 mV/km recorded before the Great Kanto Earthquake in 1923, was measured in a similar way, about 100 km away from the epicenter.

Miyakoshi (1986) observed potential differences between two parallel dipoles of unequal length, and obtained large signals in the shorter dipole (which is unusual), both before and at the time of an M5.6 earthquake (focal depth 17 km, epicenter distance 3 km). However, no precursory signal was measured before the Kobe Earthquake at Yamazaki, 25 km away from the epicenter.

Other TC observation stations detected noise in the telluric current just before the arrival of seismic P-waves in the Kobe Earthquake, suggesting the existence of a coseismic EM signal, though this EM noise might have been propagated through the atmosphere by sudden loss of electricity in the Kobe area. In the Izmit Earthquake small EM signals were recorded less than one second before the large signal associated with the arrival of seismic waves (Honkura *et al.*, 2002).

11.2.4 The VAN method

A more sophisticated form of electrode measurement, the VAN method, was developed by three Greek physicists. In the VAN method, interference from sources other than earthquakes is ingeniously screened out. VAN is an acronym from the first letters of each of the physicists' names: P. Varotsos, K. Alexopoulos and K. Nimicos (Varotsos and Alexopoulos, 1986). Two different orientations and two different electrode separations, one short (10-200 m), and one long (1-3 km), allow removal of electrical noise inconsistent with the electrode distances. Noises close to the electrodes caused by e.g. electrode instability, rainfall and other nearby artifacts can be removed using carefully calculated arrangements of multiple dipoles.

(a) Selection of suitable measurement site: Heterogeneous conductivity

There are specific sites where the measurement of Seismic Electric Signals (SES) using the VAN method can be made more sensitively than others. These sensitive sites seem to be close to geological faults composed of heterogeneous geological structures having different electrical conductivities. Suitable site selection has been discussed assuming a conductive path from the focus (Varotsos *et al.*) and has been demonstrated in a model experiment using earth waveguides with appropriate scaling (Ikeya *et al.*, 2002b, Sato *et al.*, 2001), as described in Chapter 3.

(b) VAN method controversial

The VAN method is intriguing enough to have created an impact in seismological circles and to merit further investigation, and there are enthusiastic endorsements of its predictive power in some quarters. But others argue that the VAN method has no clear mechanism. Actually, there are too many conceivable mechanisms that might account for its results. For instance, VAN scientists found that SES intensity was reciprocally proportional to the epicenter distance of the earthquakes that followed their observations. But their empirical equations, including VAN scaling laws, can also be derived theoretically as an effect of an electric field of EM waves at ULF, rather than electrostatic potential from a large dipole (Ikeya *et al.*, 1997a, 2002c). Varotsos and Alexopoulus (1986) explain their results by a ferroelectric phase transition of electric dipolar defects (orientation of dipoles in one direction). However, the Curie temperature below which ferroelectric transitions occur, is extremely low. For the dipole to carry a much larger charge than that of an electron the Curie temperature must be well above room temperature. Even so, a tiny proportion of released piezo-compensating charges in quartz grains under seismic stress might still preferentially orient themselves as ferroelectric dipoles before they disappear (Ikeya, 2000). However, because of this conflict, another mechanism, an electrokinetic mechanism involving water flow under a stress gradient, seems more acceptable to many geophysicists (Park, 1996).

Nagao *et al.*, (1993) and Uyeda *et al.* (2000) have extensively researched the VAN method and applied it, and other methods, in Japan, under the RIKEN Frontier Project. Some of their results are shown in Figure 11.3. According to the authors, spurious SES signals from the operation of electric trains are visible on the left of the figure, a genuine SES signal is shown under the pointer, and the precursory upward shift in the data line is clearly observable at night (when human activity is low). For this reason observations of most SEMS are generally made between midnight to early morning to avoid noises associated with DC-driven electric trains, engine plug discharges, industrial activities and so on. Recorded electric signals

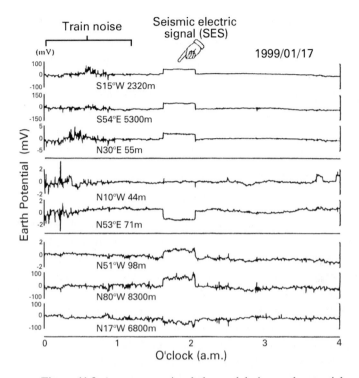

Figure 11.3 A precursory signal observed during earth potential (SES) measurements at Shimidzu, Shizuoka, Japan, using the VAN method. The directions and separations of the long and short electrodes (to remove local noise) are indicated in the figure (Uyeda *et al.*, 2000). A M4.7 earthquake occurred 11 days later at an average distance of 70 km from the three sites.

tend to reflect the seasonal, monthly, weekly, daily and holiday season patterns of human activities. Lightning also affects the signal.

The abrupt lift and fall in the base line over about 20 minutes (shown as SES in the figure) could be explained as a change in the path of the current flow as a result of fluid movement created by stress changes or underground fracture. Crystal structure changes may also explain such a change.

Preseismic SES have been observed using the VAN method but there have been few coseismic reports. This may be due to an increase in conductivity at the epicenter area after pores formed by stress-formed microfractures are filled with ground water (neutralizing the charges and diminishing the current). The lubricating effect of the ground water then triggers the fault, i.e. the earthquake occurs.

Forecasting using the VAN method has been the subject of heated debate within the seismological community over the statistical probability that VAN results are no better than coincidence (Lighthill, 1996). The controversy will obviously continue until their predictions are unambiguously successful to most scientists. In the meantime alleged statistical arguments for or against the VAN method tend to reflect how open-ended the forecast has been.

11.3   Seismo-Electromagnetic Signals

11.3.1   Precursory ULF before major earthquakes

Observation of EM noise at ultra low frequency (ULF)

Seismo-electromagnetic signals have been observed over a wide frequency range from ULF to VHF (see Chapter 3). Low-frequency seismogenic EM emissions have been observed as precursors to earthquakes and volcanic eruptions during measurements of background EM radiation (Gokhberg *et al.,* 1882; Parrots and Johnston, 1989; Yoshino, 1991). The first observation of low frequency EM emissions in Japan was a chance event during an exercise by Yoshino to measure background EM radiation (Gokhberg *et al.,* 1982).

EM waves at low frequency can travel a long distance in a conductive earth, as the skin depth for travel is large at low frequencies. Naturally enough, during the Cold War the former Soviet Union and the USA developed the use of the ULF band for military submarine telecommunications and, not surprisingly, it was during routine military communications seven kilometers from the epicenter that seismic ULF emissions were detected by electrical engineers before the Loma Prieta Earthquake (M 7.1) near San Francisco (Fraser-Smith *et al.,* 1990, Fenoglio *et al.,* 1995). (Figure 11.4.)

The characteristics of the EM field variation at ULF, 7 km from the epicenter of the Loma Prieta were:

- An increase in emissions starting 12 days before the earthquake
- A broad peak about 7 to 12 days before and then a decrease
- A huge increase a few hours before the main shock
- Continuing high levels after the main shock and during the aftershock periods.

No signal was observed at Stanford, 60 km from the epicenter and no ULF emissions were detected for the Northridge Earthquake (M6.7) at a similar military observation site more than 100 km away from the epicenter.

In what we are about to discuss the peculiar units of $nT/Hz^{1/2}$ and $V/mHz^{1/2}$ are used. These arise for both electric and magnetic fields of EM waves since we are

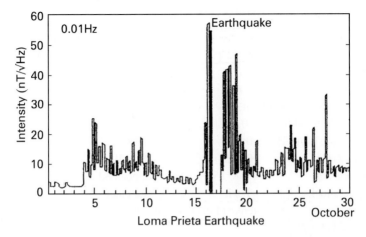

Figure 11.4 Variation of the magnetic field intensity at 0.01 Hz during October, 1989; the Loma Prieta Earthquake (M7.1) was October 17 (Fraser-Smith *et al.*, 1990). The elevated broad plateau 7-12 days before the quake and the sudden sharp rise only hours beforehand resembles the pattern of some other anomalies before quakes.

discussing the power density, which is proportional to the square of the field intensity, in units of W/m^2Hz i.e. in 1 Hz, as introduced in Chapter 3, Section 3.3.2. The observed intensities of the ULF magnetic fields—25 $nT/Hz^{1/2}$ about eight days before and 60 $nT/Hz^{1/2}$ a few hours before the Loma Prieta Earthquake—correspond respectively to an electric field of 7.5 and 18 $V/m\ Hz^{1/2}$ (a power of about 0.14 and 1 W/m^2Hz). If the intensity were really 60 $nT/Hz^{1/2}$ (18 $V/m\ Hz^{1/2}$ and about 1 W/m^2Hz), a pulse composed of EM waves of only 100 frequencies would be about 100 W/m^2 or 600 nT (180 V/m). If the pulse width were longer than 0.1 ms, an electric field of that intensity would be sufficiently high to cause unusual animal behavior.

A burst of ULF emissions was observed 3-4 days before, and 16 hours before, the Spitak Earthquake (M6.9; 25,000 casualties) in Armenia, on December 7, 1988 (Kopytenko *et al.*, 1993, Gokhberg *et al.*, 1995). Similar ULF emissions were also reported before the earthquake in Guam (Hayakawa *et al.*, 1996) as given in Table 11.1. Analysis of ULF emission data for the Taiwan-921 Earthquake revealed precursory ULF signals (Akinaga *et al*, 2001). Three magnetic field components and two electric field (earth potential difference) components indicated a relationship between epicenter distance and magnitude for ULF anomalies in the case of moderate sized earthquakes (Hattori *et al.*, 2002).

Table 11.1 Magnetic field anomalies at ULF for major earthquakes.

Place	Spitak	Loma Prieta	Guam
Date	88. 12.7	89. 10. 17	93. 8.7
Magnitude M	6.9	7.1	8.0
Depth d (km)	6	15	60
Distance R (km)	130	7	65
Magnetic field B (nT/Hz$^{1/2}$)*	0.12	60?	-
Electric field E (V/m Hz$^{1/2}$)	0.036	18?	-
Frequency f (Hz)	0.005	0.1 - 10	0.05
Time before EQ	3-4 d & 16h	Start 12d, peak 8d & 3h	14d-4d?
Before aftershocks	1-4 h	Not clear	Not clear
Reference	Kopytenko (1991)	Fraser-Smith (1990)	Hayakawa (1996)

* $E = cB = 3 \times 10^8 \, B$

11.3.2 SEMS observed before the Kobe Earthquake

(a) ELF (~300 Hz)

Preseismic SEMS at extremely low frequency (ELF) were measured at 223 Hz. At this frequency noises from artifacts and EM environments are small. Data recorded before earthquakes of M4.9 and M5.8 are shown in Figure 11.5 (Hata *et al.*, 1998). Data recorded before the Kobe Earthquake are shown in Figure 11.8.

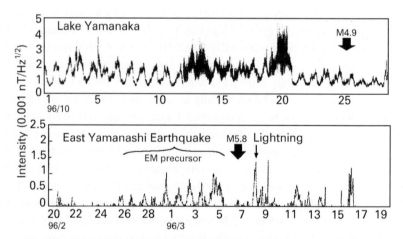

Figure 11.5 ELF emissions (223 Hz) (a) At Lake Yamanaka, 10 km from the M 4.9 event and (b) for a M 5.8 earthquake 50 km away. Daytime noise was removed by normalizing the data using daily variations at four other stations (Hata, private communication).

(b) LF (3-30 kHz) & VLF (300 Hz-3 kHz)

Fujinawa *et al.*, (1997) located sources of VLF signals before a M5.8 earthquake by monitoring EM pulses at four stations using high precision global positioning system (GPS) clocks originally made for locating lightning (Figure 11.6). However 99% of the EM pulses, initially thought to be preseismic signals, were found to be associated with lightning. Increased VLF noise was detected before the Kobe Earthquake during measurement of electric field variation using a borehole antenna (Fujinawa and Takahashi, 1995), [Figure 11.8 (2)] and increased LF and VLF noise was also independently detected before the Kobe Earthquake between January 9-11 (Figure 11.8 (3,4). However, these anomalies also corresponded to lightning, this time 150 km away over the Sea of Japan, usually too weak to be detected (Yamada and Oike, 1996). [When the recorded wave forms were analyzed they were found to be of the same form as lightning (Izutsu and Oike, 2003).] It is possible the LF and VLF waves were not generated at the epicenter but were propagated by the lightning episode and reflected back to earth more strongly than usual off a lower than normal ionosphere (diagrammatically represented in Figure 11.7) drawn down by an intense ground charge over the epicenter area, or by a related change in atmospheric conductivity which diffracted EM waves—phenomena that have been speculated about epicenter areas.

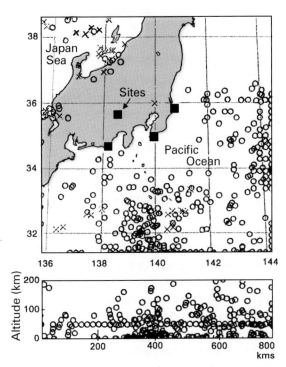

Figure 11.6 Determination of the source of VLF signals from data recorded at four observation sites, using GPS clocks for accurate times. The sites and altitudes of sources are shown. (Fujinawa *et al.*, 1997). Altitudes higher than 60 km might represent signals reflected from the ionosphere, which could include SEMS from lithosphere.

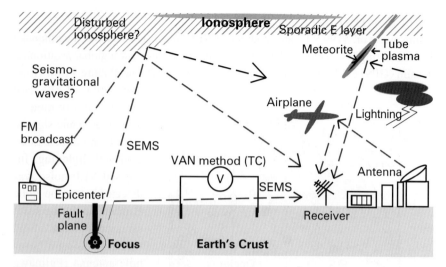

Figure 11.7 Diagrammatic depiction of various SEMS observed before the Kobe Earthquake in 1995. There may be a lithosphere-atmosphere-ionosphere (LAI) coupling before earthquakes. If plants and animals are responding to SEMS then the biosphere should be included in the coupling: (LBAI).

Franklin Japan is using a radar-technique to detect EM pulses from lightning and selling the information to the Meteorological Agency and electric power companies. The company recorded rare unidentified EM noise in the LF and ULF bands, from which noise created by lightning had been deleted, apparently emanating from the Izu Peninsula. Earthquake swarms later occurred there. (How much later is confidential information.)

Similar measurement systems have been set up in Japan by venture capitalists to expressly measure EM precursors. One business group uses a special-purpose AM radio set from which broadcast AM waves are subtracted to leave only EM noise which is then sold as noise information. Amateur hobby groups attempt earthquake prediction in Japan using their earthquake prediction apparatus: a six-band radio. Several high school teachers and groups of engineers are monitoring EM noise with various devices and publicizing the data on their Internet Homepage. EM noise from ionospheric disturbances due to solar flares or formation of sporadic E layers in the ionosphere can be verified by checking data on the Postal Ministry's Internet site.

11.3.3 Ionosphere and VLF anomalies (Kobe Earthquake)

(a) Anomalous transmission of EM waves over focal regions.

Transmission anomalies in the Omega Band (VLF waves used in navigation) were observed before the Kobe Earthquake (Hayakawa *et al.*, 1996). Transmission using

Omega waves (in a narrow band in the region of 10 - 13 kHz) has a characteristic day and night pattern, due to a day-night change in the altitude of the ionosphere, which alters the cut-off frequency of the earth's atmospheric waveguide (Section 3.3.6). Before the Kobe Earthquake there was an apparent increase in the "ionospheric daytime" in transmissions around the region. In other words the day pattern of transmission lasted longer than usual, suggesting a lower than normal descent of the ionosphere during the daytime. It is not clear whether the ionosphere does in fact descend or only appears to because of changes in near-ground atmospheric conductivity, that is, in the dielectric constant (the amount EM radiation is reflected or absorbed) and, so, in the refraction index (the amount radio waves bend at different frequencies).

From a radio transmission viewpoint, daytime seemed to be longer than the monthly average by 30 to 40 minutes. A few days before the Kobe Earthquake the variation was well above twice the standard deviation, and continued so for a period of about 10 days after the shock (See Figure 11.8 for a record of the data). A similar anomaly was also detected before the Taiwan-921 earthquake (Chen *et al.*, 2000).

To explain the phenomenon a mechanism of precursory acoustic emissions or gravity waves, which disturb the ionosphere and also generate VLF, was proposed by Molchanov and Hayakawa (1998) from studies of an additional 10 earthquakes of M>6.

(b) FM (VHF) reflected radio echo (Kushida's method)

The E-layer is an ionospheric layer that sporadically appears in spring and summer, but usually not in winter. It reflects particularly medium and short waves, but also VHF waves (30 -300 MHz). Although the Kobe Earthquake occurred in the middle of winter, wireless amateurs still found they were able to communicate long distances at 13.30 h the day before the Kobe Earthquake, using 50 MHz at 10 W.

Although FM radio signals cannot be received from beyond line-of-sight, FM radio echo from such stations can often be detected by reflection from the E-layer or from meteor trails in the atmosphere as illustrated in Figure 11.7. Two days before the Kobe Earthquake two researchers, Kushida and Kushida (1998) at the Yatsugatake South Base Astronomical Observatory, were observing meteorite trails. The radio echoes appeared as spikes on a signal base line, but the base line was fluctuating much more than normal. In fact, the radio echo became difficult to detect due to the extraordinarily large fluctuation in the base line. Two days before the Kobe Earthquake the base line significantly peaked at night (See Figure 11.8 (8). It returned to normal 20 days after the earthquake.

Curious at what they had found the Kushidas refined a method of detecting baseline anomalies in FM radio echo by orienting Yagi-Uda antennae (used in ordi-

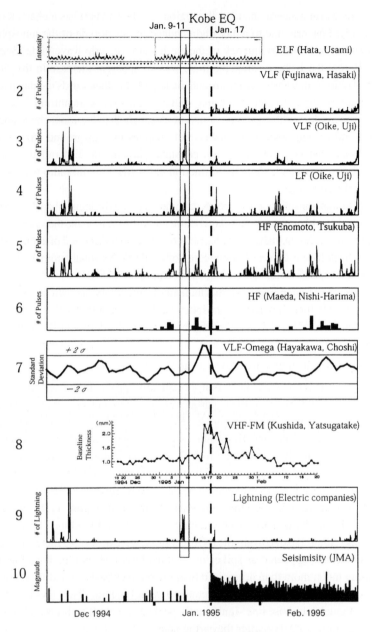

Figure 11.8 Electromagnetic anomalies at different frequencies before the Kobe Earthquake (Nagao *et al.*, 2002).

nary TV receivers) in different directions. They believed the disturbed transmission of broadcast FM waves before the earthquake resulted from disturbances in the ionosphere above the epicenter. By comparing data on base line fluctuation and behavior with known earthquakes, the Kushidas derived formulae to determine the direction of an earthquake epicenter and its magnitude. Their short-term forecasts are sold to interested supporters. Some believe the Kushida's predictions are no better than random, although there does appear to be evidence of successful prediction, except for magnitude. Kushida wrote a book in Japanese describing the suite of formulae developed to explain the relationship between baseline changes, epicenter direction and magnitude and variations between inland and shallow sea earthquakes.

Although the Kushidas see it differently, their equations can also be derived theoretically from the skin depths of ULF waves (rather than VHF waves) emanating from the earthquake focus through seawater and the Earth's crust. Preseismic EM waves at ULF may account for their observation of disturbed VHF waves. That is, the ionosphere might be affected by ULF waves due to an EM lithosphere-atmosphere-ionosphere coupling (LAI), either as a result of conductivity changes in the atmosphere above the epicenter area (due possibly to ionization by emanated radioactive radon), or to the formation of an intense ULF electric field in the epicenter area, disturbing the propagation of FM radio waves (as if the local ionosphere had descended above the epicenter area).

Another mechanism (as touched upon in Section 11.3.3 a) has been proposed to explain LAI coupling: a dynamic coupling of acoustic waves in the atmosphere, sometimes called "internal gravitational waves". In this mechanism the descent of the ionosphere is caused by gravitational vibration created by stress-induced rock fractures in the lithosphere (preceding earthquakes) thus allowing the over-horizon reception of FM radio waves (Pilipenko *et al.*, 2001).

(c) Coseismic VHF signal near the observation site?

Yoshida and Nishi (2001), telecommunication engineers, claim to have detected coseismic emission of VHF waves. But the skin depth of EM waves at VHF is small, meaning they should not be able to reach the earth's surface. This raises the possibility that the noise detected in the VHF band was actually created by a combination of other frequencies in SEMS pulses. Observing the VHF band Yoshida and Nishi also detected galactic noise, meteor showers, solar flares and human EM artifacts from 86.7 MHz with sensitivity down to -120 dBm, and the aforementioned coseismic VHF signal of - 90 dBm, whose source seemed close to the monitoring site. These researchers have not detected a clear VHF precursor signal so far.

11.3.4 Observation of SEMS by satellite

(a) ELF from the satellite Cosmos

One and a half to two and a half months after the earthquake at Spitak, Armenia (M 6.9 December 7, 1988), the low orbiting Russian satellite Cosmos-1809 observed bursts of ELF emissions for one to three minutes above the aftershock area (Serebryakova *et al.*, 1992). Similar EM bursts were also observed in the region on two frequency bands, ULF-ELF (<1 kHz) and VLF (10-15 kHz) during 180 orbits of the Intercosmos-24 satellite from November 16 to December 31, 1989, nearly above the epicenters of twenty-eight strong earthquakes (5.2 < M < 6.1). The ULF-ELF spectra showed greater intensity the lower the frequency (i.e. A plot of the ULF-ELF spectra showed more ULF waves than ELF waves). Far from the Spitak epicenter VLF emissions were observed 12-24 hours before the main shock (Molchanov *et al.*, 1993).

(b) Pioneering projects and proposals

In 1996, NASDA (Professor M. Hayakawa, Telecommunications Engineering), set up the Earthquake Remote Sensing Frontier Research Project in Japan to study SEMS phenomena in terms of a linked lithosphere, atmosphere and ionosphere. The project's final report was produced in 2001 and the initiative gave rise to a series of symposia whose proceedings were published (Hayakawa and Molchanov, ed. 2002). These report that the Intercosmos-24 and Cosmos-900 satellites observed reliable correlations between plasma density variations in the ionosphere and global earthquake distribution. The findings were true for daytime, calm geomagnetic conditions, at an altitude range of 500-800 km and 50 degrees latitude.

In addition to observing EM phenomena, satellites have been used to investigate relationships between acoustic emissions and temperature anomalies. Although not published, some of the results have been communicated personally by US-NASA researchers. The objective, if possible, is to allow the public to tune in to earthquake forecasts in much the same way they tune in to weather forecasts.

Another project, Frontier RIKEN (organized by geophysicists S. Uyeda and T. Nagao), set up an interdisciplinary working group within the International Union of Geodesy and Geophysics (IUGG) to look at electromagnetism, earthquakes and volcanoes. The group organized several special sessions and symposia, and recently dedicated a special issue of the Journal of Geodynamics to recent investigations of earthquake-related EM (Uyeda and Park, ed. 2002).

A large number of papers have been published in this emerging field in these last two decades coincident with technological innovations in EM detection and better information processing. If the continuing research advance is able to clarify reasonable EM processes in the Earth's crust preceding earthquakes, measurements of the process might be able to be used to anticipate shocks.

Electromagnetic Seismology (the author's term for the study of SEMS and other precursory phenomena) is another pioneering approach in which rapid advances have been made in the eight years since the Kobe Earthquake. Although short term earthquake prediction is not straightforward and, according to its strongest critics, may not even be possible from EM emissions from microfractures, real time observation of underground fracture events may nevertheless provide enough hints of an impending major quake that lives will be saved.

(c) DEMETER satellite to observe Earth's EM environment (Detection of EM emissions from earthquake regions)

Some time in 2004 a microsatellite will be launched by the French Centre National d'Etudes Spatiales (CNES). The major scientific objectives of the DEMETER mission are to obtain global information on the Earth's EM environment and disturbances at the satellite altitude due to human activities and pre-/post-seismic

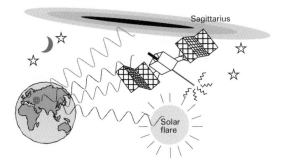

Figure 11.9 Satellite observation of the Earth's EM environment to detect SEMS.

activities (Figure 11.9). Space scientists hope to understand the mechanisms of the latter (Parrot, 2002). If a descending ionosphere over detectable SEMS from an epicenter area before a large quake were observed it would be a catalyst for a research breakthrough in prediction of large quakes. Although lightning near the ground and above the ionosphere (sprites) may disguise or disturb EM signals from underground, DEMETER promises to be a useful source of research information.

(d) Cosmic Ray proton effect before Kobe Quake

A satellite monitoring cosmic ray protons recorded an anomalous decrease over the SE Asian region before the Kobe Earthquake. The cause of this is not certain, but as shown by Tokyo neutron monitor data, a local decrease caused by the Earth's magnetic field (a phenomenon called the "Forbush decrease") was not in effect at the time. It may or may not have been related to the Kobe Earthquake. Further research into possible effects of SEMS at ULF on ionospheric cosmic ray proton activity before large earthquakes would be helpful.

11.3.5 Recording preseismic EM pulses with a Pulse Height Analyzer

As mentioned, EM waves are emitted in a wide range of frequencies from DC to VHF. Usually, detection over a narrow band is preferable to remove noise. This implies the use of various detectors each operating in a narrow band. The low frequency component of EM pulses would be measurable as ULF or even DC earth potential changes. Such pulse-like geoelectric signals have been measured using a vertical dipole antenna buried under the ground to eliminate industrial and meteorological electrical noise (Enomoto *et al.*, 1997).

This laboratory has employed a pulse height analyzer (PHA) to record pulses of greater intensity than those produced by normal city activity (Yamanaka *et al.*, 2002). A PHA is widely used in nuclear science and engineering to identify radiation energy and radioactive nuclei. Although it is possible to buy dedicated PHA instruments, it is often simpler today to convert a personal computer into a PHA by inserting a PHA board. The PHA must be able to measure positive and negative pulses since SEMS might be either. Figure 11.10 shows a PHA record for the month of May 2000. An anomalous EM pulse height distribution occurred on two

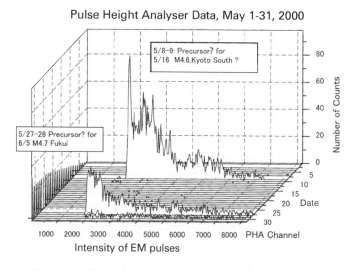

Figure 11.10 The number of EM pulses (time by antenna) stored by a Pulse Height Analyser over the month of May 2000. Anomalous pulses appear on May 7-8 and 27-28 with average daily readings in between. Intense pulses are stored at channel 3000 and moderate ones around channel 4800. Nearby earthquakes that occurred about 8 days later are noted in both cases, with their magnitudes (Yamanaka *et al.*, 2002). (The abrupt cut-off in the data on the left is intentional, representing removal of environmental EM noise.)

days in the month, compared with average days between. The spectra for May 8-9 occurred about eight days before a M4.6 earthquake (epicenter distance 20 km, focal depth, 10km); the spectra for May 27-28 about eight days before an earthquake of M4.7 (distance 120 km, focal depth 10km).

11.3.6 Real-time monitoring of SEMS

If SEMS are detected earlier than seismic P-waves, they can be used as a primary signal of earthquakes. An early-warning system might be set up around electronic detection of EM events before the P-wave—the exact triangulation of time and place of pulse origins determined by observation at several sites, using GPS clocks to better than a microsecond precision (which would give location to within 300 m in distance). This sort of real time monitoring as an alarm of an impending earthquake (rather than a prediction of magnitude, epicenter and time) would still be useful.

Measurements of broadband EM waves and their frequency components may allow us to estimate the shape of the EM pulses. Electric parameters such as the dielectric constant and resistivity of the underground crust would be deduced from this by on-site computer analysis of their shape.

11.3.7 Earth-origin electric pulses and Earth crust waveguide

There are difficulties in distinguishing SEMS from atmospheric EM waves arising from lightning, or magnetic storms caused by solar flares, or galaxy EM waves or the enormous EM artifacts in modern cities. To eliminate as much extraneous noise as possible it is best to measure close to the source at a place where noise is lowest. Seismic EM signals from underground are best monitored in boreholes underground or on the deep sea floor. Wet sediments and seawater also act as shields against atmospheric EM noise, human EM artifacts and atmospheric lightning.

Bore-hole:

Tsutsui (2002) placed two sensors: one 5 m above the ground and the other 90 m underground in a bore-hole measuring 10 cm in diameter [See Figure 11.11 (a)]. One month before the Western Tottori Earthquake (M7.2) at 13.30 h on October 6, 2000, Tsutsui detected earth-origin EM pulses. The site was 200 km from the epicenter, and linking the pulses with the earthquake was difficult, as usual.

Dynamic frequency spectra:

(a) Evidence of earth-origin EM pulses

Figure 11.11 (b) shows swarms of electric pulses recorded by the two sensors above and under ground. Intense electric pulses in the frequency range 300 - 700 Hz were observed more clearly in the bore-hole than in the atmosphere, suggesting EM waves in these frequencies were dissipated by the wet conductive soil before they reached the detector in the atmosphere.

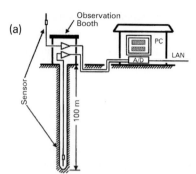

Figure 11.11 (a) Configuration of a system for measuring the underground EM environment. (b) Dynamic frequency spectra exhibiting definite evidence of earth-origin electric pulse swarms in the earth (lower spectra) and leakage from the ground (upper spectra). The intensity above the ground is less than underground by -14 dB (Tsustsui, 2002).

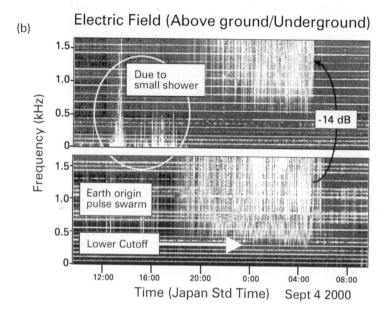

(b) Evidence of the Earth-crust waveguide: Moho or Conrad Plane?

There were clear lower frequency cut-off in the borehole spectra suggesting the existence of an earth-crust waveguide as proposed and modeled experimentally (See Chapter 3 and Ikeya *et al.*,1997, 2002). An earth crust waveguide was described for the propagation of EM waves, in addition to the atmospheric waveguide formed by the ionosphere and the surface of the earth. The underground conductive layer was estimated to be 250 - 160 km deep using a specific dielectric constant ($\varepsilon^* = 9$) and attributing it to the Moho plane (the boundary between the lower crust and the mantle). However, if a low frequency dielectric constant ($\varepsilon^* = 1000$) of the mineral-water boundary at 10 kHz were used, the depth of the boundary reduces to 20-10 km, exactly the depth of the Conrad Plane, the boundary between the upper and lower crusts (Ikeya *et al.*, 2002). The cut-off frequency at low frequency considering the large ε^* would be about 10 Hz. Hence, most EM waves would be propagated through earth's dielectric waveguide.

11.3.8 Electric field intensity differences between VAN and SEMS: Is $\varepsilon^* = 10,000,000$ at 1 Hz?

One might wonder why the electric fields obtained by EM pulse measurements and those derived from the field intensity of ULF waves in the atmosphere are intense while the earth potential (as measured by the TC or VAN method) is only of the order of 10 mV/km (Section 11.2.3) or 100 mV/km at the most. This is six orders of magnitude smaller than the 0.3 - 30 V/m calculated from the change in the magnetic field intensity of 1 - 10 nT before large earthquakes.

One explanation is that SEMS comprise a range of frequencies, meaning that the total energy is thousands or a million times greater than the conventional measurement of SEMS within narrow bands of 1 Hz. If the SEMS is a pulse composed of EM waves at different frequencies (like the wave packet shown in Figure 3.9), the peak intensity of the pulse could be several orders of magnitude higher than a few V/m Hz$^{1/2}$ taking into account the frequency distribution of the EM waves.

A further explanation is that the specific dielectric constant, ε^*, of the crust or the surface wet soil at ULF is not $\varepsilon^* = 8$ as found for granite (Section 3.3.4 and Table 3.4), but might be tens of millions, as discussed for a decay time of 4.3 seconds for charges induced by stress release (Sasaoka, 1998). This leads to enormous differences in measured voltage depending on the precise angle of detection and other local factors, and could account for VAN/ULF differences. Recent data (Hattori, 2003 private communication) support a large ε^*, but simultaneous absolute measurements of ULF magnetic fields and earth potential difference would need to be made at several sites to help clarify the picture.

The extraordinarily large ε^* at low frequency and the change in ε^* with frequency might be the key linking SEMS with unusual animal behavior. The electric and magnetic fields of an EM pulse having a peak energy of 1 W/m^2 in air at 1 Hz are tabulated with their velocity in Table 11.2 with the electric field $1/\varepsilon^*$ times lower ($1/\varepsilon^{*0.25}$ times given the impedance of EM waves). Measurements of magnetic fields at 80 Hz were easier than that of electric fields across an underground tunnel (Yoshino and Sato, 1998), presumably because of the extremely small electric fields due to propagation in a conductive medium with a high apparent ε^* at that low frequency.

Table 11.2 Electric fields of SEMS with 1 W/m² at less than 1 Hz in various media.

Media	(Dielectric constant)	E (E of EM waves)[3]	Magnetic field	Velocity
Air	($\varepsilon^* = 1$)	20 (20) V/m	67 nT	300,000 km/s
Granite	($\varepsilon^* = 8$)	2.5 (12) V/m	67 nT	100,000 km/s
Water	($\varepsilon^* = 80$)	0.25 (6.7) V/m	67 nT	33,000 km/s
Crust & soil[1]	($\varepsilon^* = 10,000,000$)	2 (180) mV/km	67 nT	100 km/s
ACROSS[2]	(300 Hz: $\varepsilon^* = 160,000$)	0.12 (1) V/m	67 nT	750 km/s
	(10 kHz: $\varepsilon^* = 10,000$)	0.02 (2) V/m	67 nT	10,000 km/s

[1] See Table 3.4 and Section 3.3.4. Only sensitive animals respond to this intensity.
[2] ε^* was calculated from the experimental propagation velocity of EM fields (Kumazawa).
[3] Electric fields with intensity proportional to $1/\varepsilon^*$ (and to $1/\varepsilon^{*\,0.25}$ using the impedance).

The electric field intensity and speed, c^* of EM waves (100 km/s, and so a wavelength of 100 km), become very low compared with the speed of light. Recently $c^* = 750$ km/s (leading to an apparent $\varepsilon^* = 160,000$) was observed in the EM-ACROSS experiment at 300 Hz (T. Nakajima *et al.*, 2003). This means coseismic EM signals may be used to create an earlier warning system than those using seismic P-waves (see Section 3.2.2). This hypothetical treatment of the dielectric constant explains what specialists call "diffusion fields" in a conductive media as fields of apparent EM waves with an extraordinarily large ε^* and slow velocity, c^*.

11.3.9 Empirical scaling relations explained as EM waves

(a) Empirical relations derived theoretically using the EM model of a fault

Figure 11.12 shows unusual animal behavior as a function of distance from the epicenter and the magnitude of the earthquake (M). The figure is from Rikitake (2001). The right-hand cutoff in the data was also seen when many other precursors were similarly plotted. Rikitake inserted the dashed line there and its equation is shown in the figure, representing the most sensitive animal response with distance and magnitude of the event.

Using the EM model of a fault in which EM waves are produced by piezo-compensating charges (Section 7.4.1 and Ikeya *et al.*, 1997a, 2000), it is possible to derive Rikitake's empirical line theoretically and to calculate the voltages which would be produced at various distances from the epicenter and at various magnitudes. The model assumes that the volume of the high stress area is proportional to the cube or square of the fault length for small or large earthquakes respectively. It also assumes a piezoelectric constant of quartz (10^{-12} C·m) and that 10% of the rock is quartz completely devoid of crystal orientation that would increase the voltages. The theoretical line produced is the solid (bent) line to the right of the Figure which is only 3% different in slope from Rikitake's empirical one, hence remarkably close.

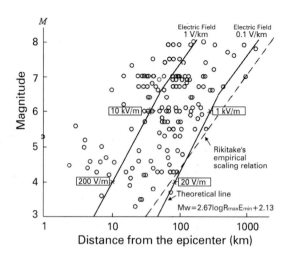

Figure 11.12 The relation between magnitude and epicenter distance for animal behavior explained by the EM model of a fault. The voltages in the boxes assume a 1% initial orientation of quartz grains and orientation of piezo-compensating charges. (For explanation of the theoretical lines, see text.)

The absolute values of the voltages produced are quite low and even for M8 at the top of the line the indicated field is only 0.1 V/km (= 0.0001 V/m) and decreases with lesser magnitude events. The sensitive catfish could respond to a value like that, (0.005 mV/m) but more typically this voltage is about 1000 times less than those which are shown in *Earthquakes and Animals* to cause unusual behaviors in many species. This low value applies about 100 km from the epicenter, but the voltage is only about 10 times higher when the epicenter distance is about 10 km (indicated by the bent solid line to the left).

However if there is even a small amount of orientation in the quartz crystals the picture changes very dramatically. The boxed voltages on the two solid lines are the levels created if there is only 1% quartz crystal orientation: for magnitude 6 and 100 km the voltage of ca. 1000 V/m is so high that it would be likely to stun or paralyse animals in the area. (In fact, there are many reports of dead animals at the time of large earthquakes.) This calculation means that even if the quartz orientation is a very minimal, 0.01-0.1%, it would still create the kind of voltages which animals have been shown to respond to in experiments described in *Earthquakes and Animals*. This shows that the model contains plenty of latitude for producing reasonable voltages at reasonable distances from the epicenter.

(b) Precursor times derived theoretically from a dilatancy model of a fault

Rikitake's empirical scaling relations for magnitude and precursor time (1992) can also be produced for unusual animal behavior (Ikeya *et al.*, 2000) using a model based on a dilatancy model of a fault (Schultz *et al.*, 1973), (Section 3.2.4). The list reflects an increase in scale.

- Microfractures, which produce dilatancy, generate EM pulses (and acoustic emissions). Both are attenuated during propagation, but EM waves at ULF (and those at some higher frequencies produced by resulting voltage discharge in the atmosphere) are detected by sensitive animals.
- A series of microfractures in large asperities (large rough areas of rock) create electric charge and unusual phenomena in the epicenter area.
- Fluid diffuses gradually into microfractures.
- The microfractured region becomes fragile and cannot withstand the seismic stress.
- Asperities blocking movement of the active fault, fracture.
- The fault moves, leading to a large earthquake, if not blocked by other asperities.

An earthquake occurs when the entire microfracture volume (the size of which dictates the magnitude of the earthquake) is filled with groundwater, as schematically illustrated in Figure 2.17. Hence, the precursor time is the time required to fill the volume with ground water. The empirical scaling laws of precursory time as well as others obtained by Rikitake (2000) can be derived theoretically based on this model.

A precursory time of 8 days for an M7 inland earthquake would be the time required for the diffusion of water into about a 10 km x 10 km area of the fault plane, hence lubricating it and perhaps triggering earth movement. It becomes 5 to 7 days for an M5-M6 and 3 to 4 days for an M4 earthquake, according to our observation, depending on the relation of earthquake volume to magnitude. However, although a certain precursor may suggest a certain number of days before an earthquake, it is more important to know the precursor intensity and the area over which it appears. Observations from number of stations will help derive a magnitude for the quake.

Piezo-compensating charges produced by randomly oriented quartz grains gave EM pulses three orders of magnitude too small to explain animal phenomena, but, a mere 1 % alignment of piezo-compensating charges does give a sufficiently intense electric field. Frictional electricity or contact potential of large and fine fragments of microfractures may also play a major role. Other mechanisms associated with electrokinetics of fluid and ferroelectric orientation of large charge dipoles might also be operative underground. The mechanism of EM generation by microfracture awaits clarification by both theoretical and experimental physicists.

All earthquake prediction studies using EM waves are at a premature stage and embroiled in scientific controversies. The matter will only be resolved by the observation of an intense pulsed electric field in the epicenter and surrounding areas.

11.4 Summary

Researchers in an increasing range of disciplines have observed EM signals before earthquakes over a range of frequencies from ULF to VHF. The proliferation in research interest indicates that the field is yielding results—that there is some relationship between earthquakes and generation of EM waves —but the picture is confused in terms of mechanisms and it is very difficult at this stage to pin observation of SEMS down to specific earthquakes. Though different EM detection methods have had varying degrees of forecasting success, skeptical seismologists have challenged the results on grounds that any "successes" are no better than random and any links with earthquakes are imaginary.

Our own EM model of a fault argues particularly for EM waves at ULF frequencies reaching the surface in the epicenter area and creating intense electric fields resulting in discharges generating high frequency EM waves. These pulses are capable of causing unusual behavior in a range of animals, plants and domestic appliances, as this book has attempted to demonstrate. Difficulties remain over discrepancies between electric field intensities as measured by TC and by SEMS detectors, but these may be resolved through better understanding of the magnitude of the dielectric constant. The EM and dilatancy models of a fault can also explain the scaling relations between magnitude, distance and animal behavior that have been arrived at by Rikitake.

As in all new research fields there are pitfalls in SEMS detection, not least in ascribing an apparent precursor to an EM effect when other factors are at work. These are inevitable in interdisciplinary work at an early stage but it can be hoped that as the research advances and becomes increasingly multidisciplinary, such incidents will occur less often.

Continuing satellite observation will hopefully shed more light not just on SEMS but also on the intriguing LAI coupling (or, if plants and animals are involved, an LBAI coupling) that leads to an apparent descent of the ionosphere, perhaps over epicenter areas where ground charges build up, possibly from huge rock stresses before earthquakes.

12

Summary

Japan is the most earthquake-prone country on earth.

Many of its earthquakes are tens of kilometers deep and caused by one tectonic plate sliding under another. Many more earthquakes occur near the surface and are often very destructive. So it is natural that there should be a very active program of earthquake research to prevent as much injury and damage as possible. Because this research is particularly useful within Japan, the results do not necessarily find their way into international media. That may be why a significant proportion of the material in this book may be new to readers.

What is also unusual about Japan is the number of amateur groups involved in the research. Sometimes an expert is intimately involved with them, but some are solely self-funded citizens' initiatives. Such intense amateur involvement seems unparalleled in other countries.

12.1 EM research is still young

The research summarized in *Earthquakes and Animals* has been conducted in a University Department by a group of staff and postgraduate research students over a period of eight years. The outcome was over 40 published papers that had to pass the usual hurdles of editors and referees. This is not an amateur effort and should not be judged as one. Nevertheless, the research is young and has not gained international recognition yet.

For example, the General Assembly of the International Association of Seismology and Physics of the Earth's Interior (IASPEI) has a subcommission on earthquake prediction that has compiled a preliminary list of precursors (Wyss, 1997). It says that inclusion in the list does not make a precursor "true" and exclusion does not necessarily make it "false".

The IASPEI preliminary list of significant precursors, March, 1994, reads:

(1) Foreshocks (hours to months)
(2) Preshocks (month to years)
(3) Seismic quiescence before strong aftershocks

(4) Radon concentration and temperature decrease in ground water

(5) Ground water rise

In the second round of evaluation, Professor T. Yoshino unsuccessfully proposed EM emissions related to earthquakes in Japan as a significant precursor (Gohberg, 1982). No nomination has yet been made for precursive animal behavior, which is still considered to be far from scientific.

The nominators have to present:

(1) The probability that the anomaly and mainshock coincided by chance

(2) A plausible connection between the anomaly and the mainshock

(3) Excellent quality data.

Obviously more time, and evidence of the EM precursor hypothesis explored in *Earthquakes and Animals,* is going to be needed before sceptics are convinced and EM precursors find their way into the IASPEI Preliminary List of Significant Precursors. Perhaps the integrated science network in some Japanese High Schools, observing catfish and SEMS signals simultaneously throughout Japan, may yet corroborate our hypothesis. It's a challenge to high school teachers and pupils to prove or disprove the ancient catfish legend and to present their data to the IASPEI committee for scientific examination.

However, because EM precursor research does not yet have the backing of international organizations, many continue to regard it with skepticism. It has been called pseudo-science or even pathological science, a term coined by the Nobel Prize Laureate, Irving Langmuir, to imply poor, slipshod and misleading research. There is a persistent belief in the seismological community that there are simply no precursor phenomena before earthquakes. There is also a belief that lay citizens' observations are not an appropriate subject for research. At times the objections from the anti-prediction quarter were as emotional and unreasonable as the insistence of some lay citizens over the precursor significance of some of their unusual clouds. There have been numerous attempts to find some truth to the legends and reports of precursor phenomena before earthquakes, but the exercise has been personally too costly for many: even sympathetic professors and colleagues urged me not to ruin the prestige of my university and laboratory, and the future of graduate students, by persevering with it. We met at times, implacable and hostile resistance and contempt from funding agencies and from referees of scientific journals as we sought to get the work published. But we persisted with the research, because of what we were finding, and because of encouragement from some top-grade seismologists who believed EM fields were involved. The more constructive the criticism we received, the more exacting our experimental work became.

Nevertheless there will still be some who consider this work to be "rubbish", the inevitable outcome of research done by someone who has strayed too far from his specialist field. But, someone had to test the hypothesis—if only to show, in the end, that it was unfounded.

But we don't believe it is unfounded. In our opinion, the biological and meteorological phenomena before earthquakes reported by lay people and mentioned in legends, can be ascribed to EM pulses caused by underground rock fracture ahead of the main shock. More specifically we would attribute them primarily to EM waves (seismo-electromagnetic signals, SEMS)—at ULF frequency travelling to the surface in the epicenter area and creating strong electric fields, whose local discharges generate high frequency EM waves affecting not just animals, plants and objects, but also the atmosphere.

12.2 Responses to specific objections

Three specific objections in particular are worth particular attention.

1.   The survey material is so subject to psychological factors, that it can say nothing about the validity of precursors.

There are definitely strong psychological factors affecting reports from citizens, who will have varied motives ranging from a desire to help, to sincere but quite erroneous correlations they imagine they observe, or to malicious interference with any survey.

However it has proved possible to examine the reliability of the reports using the statistical Chi-squared test (Whitehead *et al.*, 2003) and some have been noted in this book. In general, it appears that the results are not random or entirely psychologically influenced. There are large psychological effects, yes, but also genuine reports of precursors, which are found within specific time ranges and within specific distances (tens of kilometers) from the epicenter. It is rather remarkable that a sociological method should give some useful information on whether a physical precursor may be genuine, but that appears to be the case here.

2.   The research does not observe the principles of the order of scientific investigation.

There is general agreement that scientific research should proceed according to an established hierarchy of principles: mathematics leads to physics, physics leads to chemistry, which leads to biology (See Figure 12.1). At each stage, there are emergent features which cannot be predicted from the disciplines below. Arguing back from uncertain rare phenomena in biology and meteorology to basic physics may give a radical impression of driving backwards along what is really considered to be a one-way street.

In the author's opinion this is to misunderstand the nature of investigation when a new scientific field is under study. An observation leads to an idea, which is tested, and may become a hypothesis, which is then further tested. The researcher may start at a rather odd point on the pyramid and move up and down it in a rather haphazard way at first. A

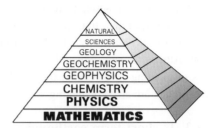

Figure 12.1 The traditional hierarchical approach to scientific research. According to the model research should build from a proven basis in mathematics rather than start at higher levels and then attempt to validate itself by working backwards.

two-way system is the natural state of affairs when a new field is being born. All established sciences have once passed through this stage.

It is true that to move from precursor reports and legends to the conclusion that, since animals respond to electric fields of EM waves in the laboratory, there must therefore also be electric fields of EM waves present round epicenters, is something akin to a leap of faith down the hierarchical pyramid. But it seems reasonable to consider this approach a step towards an answer. If the preliminary assumption is made (for experimental purposes) that the precursor statements are correct, then the experimenter may be allowed to assume so, if the physical quantity (EM waves) is later observed before earthquakes and theoretically proved to be appropriate (in terms of the EM model of a fault presented in this book).

3. Any apparently positive results cannot be correct because the formation of an earthquake is a nonlinear process and hence unpredictable in principle.

The nonlinear theory means that a very small trigger sets off fracturing and movement in the earth but it cannot, even in principle, be calculated whether the movement will develop into a full-scale earthquake or, literally, grind to a halt because of large asperities.

This places very high and probably excessive value on the nonlinear theoretical model. Is it possible to prove beyond any doubt that the process is completely chaotic? It is always possible that there is a subset which is not completely chaotic. Therefore it is reasonable to investigate some apparent correlations, particularly if there are genuine claims that precursors seem to have happened.

One of the encouraging trends to emerge during the present program of research has been that, internationally, several research groups are now taking EM precursors very seriously: The First European Workshop on Earthquake Prediction (Athens, Greece, November 5-6, 2003) featured papers on EM wave anomalies originating from China, Italy and Russia. Countries not represented at the conference are also conducting such research.

12.3 Is short-term earthquake prediction possible?

If electromagnetic radiation is emitted coseismically and can be detected essentially simultaneously, it opens up the possibility of some warning (even if only seconds) before the first P- or S- waves arrive, and that would be a revolutionary advance. If, better still, the preseismic emissions can be confirmed, warnings of hours or days may be possible.

However, most seismologists appear to have given up short-term forecasting of earthquakes to concentrate on earthquake mechanisms. We certainly need to understand the mechanics of the nucleation phase of earthquakes, but not to the extent that useful research going on internationally on EM anomalies is being disparaged, or not taken seriously.

Accurate forecasting is next to impossible, and, at the moment, the nearest seismologists can come to prediction is a calculation of increased risk based on increased seismic activity. This is better than nothing, but it may be that real time observations of a broad range of presursor EM phenomena may be more reliable than simply using increased seismic activity. Though the research is certainly not at the stage where unusual EM phenomena alone can be relied upon in any official sense, they are useful indicators to lay people (who own pets and electronic objects), and when taken together with other indicators, may prove to have significant general forecasting and life-saving value. In our own case, it is appreciated that local fractures producing electric pulses may not always lead to major earthquakes, just as increased seismic activity does not tell us whether a small earthquake is an isolated event or a foreshock of a larger quake to come.

12.4 What would be the next steps in the research program?

It would be very useful to have direct and detailed evidence that electric charge is present before earthquakes. This book has accumulated many indications that this is so (as enumerated in Chapter 11 and deduced from the rock crushing experiment in Chapter 5) and justified it on theoretical grounds, but was only able to report a few scattered observations from various countries, essentially serendipitous in nature. More evidence is necessary.

The pilot school project using catfish is a model which could be very valuable on a larger scale. Even if no earthquakes occur, it teaches students useful cultural, scientific and practical lessons. But animals require care and feeding, which may not be ideal outside the student context. It would be good to find an indicator requiring no more care than already given. Perhaps indicators from agricultural crops or animals could be developed, hence the discussion in Chapter 10 of veterinary devices, some already on the market, to monitor animal physiology. Ideally a network of recording electrometers (measuring electric charge) should cover a wide area, but this is impractical. Thought should go into a cheap indicator (at essentially zero cost) which could be mass produced and strategically placed throughout a country (perhaps in Post Offices or other widely distributed Government Buildings) so that changes in certain areas could be observed and reported.

12.5 Conclusion

My research into earthquake precursors began as a response to the Kobe Earthquake in January, 1995. I was appalled at the loss of life around me and my inability to do anything to relieve the suffering. I had three decades of scientific work behind me using Electron Spin Resonance (ESR) in the fields of solid state physics, geology and anthropology, and a desire to use that experience if possible to give forewarning of these massive forces unleashed by movements in the earth's crust.

I believe that *Earthquake and Animals* provides evidence that EM waves play an important and unfolding role in earthquakes (from the nucleation phase right through to main fracture). EM signals are being detected before earthquakes and their appearance at the surface from stressed rock underground can be accounted for theoretically. Our many experiments have demonstrated that animals, plants and objects, respond to EM waves in the laboratory much as the old legends and many modern reports say they do before earthquakes—depending on many variables: frequency, size, resistance, pulsewidth, and angle. EQCs, EQL, EQFs, and the odd appearance of the heavenly bodies can also be explained as electric field effects. The case is certainly not established, but it has a lot of circumstantial and empirical support.

The research fell between disciplines and is thus open to the criticism that it lacks specialist rigor and interpretation, but we are also scientists, not amateurs. An interdisciplinary approach was forced on us largely because most seismologists have not studied electromagnetism, and biologists know little about earthquakes. It simply illustrates the point that, as knowledge increases and researchers increasingly become isolated in their specialist fields, an interdisciplinary approach to complex events affecting human populations becomes imperative.

Our concern remains what it was at the beginning: the saving of human lives. Naturally, the exercise also became a fascinating exploration of the possibility that the earthquake legends—so common in earthquake-prone countries—are not merely superstitions but actually have a scientific basis. We believe we have gone a good distance towards establishing the latter but not so far towards the former. A seismologist may argue that SEMS precursors are minor phenomena and that even if people did observe them and make moves to protect themselves from large earthquakes, less than one percent of deaths would be avoided. But one percent would still have been 64 lives saved at Kobe, about 200 near Izmit and 2400 in the Tangshan Earthquake in China.

In fact, we are inclined to believe that, though perfect earthquake prediction (with rigid definition of magnitude, place and time) is a near impossibility, a general forewarning from intelligent analysis of precursor phenomena, combined with personal precautions, could reduce injuries and deaths to levels substantially below what could be attained with the best earthquake engineering. Such measures would be of considerable benefit in Japan, particularly, whose intricate physical and human infrastructure, is quite vulnerable to disruption. But the same principles should apply in many other seismically active areas—China, New Zealand, California, the Andes Chain, the Aegean, and parts of the Mediterranean.

The research was both exhilarating (because of what we found) and exhausting, because of the opposition we frequently encountered. But we like to think that we have unearthed a possible new discipline—a field that we have, in hope, called Electromagnetic Seismology—that links underground rock fracture, seismic EM pulses and rare natural phenomena. And it is also our hope that it may become a useful tool in forecasting the likelihood of a large earthquake, in a general vicinity, within a period of less than a few weeks.

In any event, disaster prevention of the kind outlined in Appendix 2 is a much more productive exercise in earthquake-prone countries than endless controversies over whether or not prediction is even possible.

Appendix 1

Questionnaire to Scientists and Citizens

Are earthquake precursors superstitions or real phenomena ?

Four hundred questionnaires were sent to fellow scientists and lay citizens in Japan, and to scientists in the USA and Europe, who, except for a few, have not experienced large earthquakes. The survey sought opinions about the earthquake precursors discussed in this book. Respondents could answer in one of three ways.

Such earthquake phenomena are:

(1) Superstitions or afterthoughts

(2) I'm uncertain

(3) Real phenomena

The questionnaire was sent to:

Japanese lay citizens: Return postcards to 120 citizens (60 males and 60 females) by random selection of addresses from NTT telephone directories all over Japan, taking the population distribution into account—about one person per million residents in different areas.

Japanese physicists, seismologists and zoologists: Return postcards to 100 members chosen at random from directories of the Japanese Physical Society and the Seismological Society, and 80 members of the Zoological Society.

US and European geophysicists: E-mails to 100 addresses chosen randomly from a directory of US and European geophysicists.

When a card or an e-mail was returned as undeliverable an adjacent name was chosen from the directories.

Response:

There was only a small response: 20 (12 males and 8 females) from 120 citizens; 28 from 80 zoologists; 38 from physicists and 36 from seismologists, of 100 surveyed; 15 from 100 USA and European geophysicists. Results were normalized to 100.

Analysis:

Japanese lay citizens and scientists:

Two thirds of citizens and scientists who responded believed unusual animal behaviors before earthquakes were real phenomena. Only 4 % considered it to be superstition. Less than half believed unusual plant and electric appliances were real phenomena. Ninety percent of seismologists have read books on unusual phenomena before earthquakes, but only 10 % of citizens, physicists and zoologists. Those who thought anomalous behavior of plants and electric appliances and unusual atmospheric features were real phenomena were 26 %, 19 % and 6 %, respectively. The percentages for EQL were 45 %, 45 % and 9 % for real, uncertain and superstition. The large number in the "uncertain" category could be accounted for by Japanese kindness in matters of this nature i.e. an unwillingness to discourage a struggling scientific endeavor.

USA and European geophysicists:

The small number of responses showed a general lack of interest, and, the responses themselves, a general scepticism about unusual phenomena before earthquakes. Two thirds of those who answered believed the precursor phenomena were superstitions. Only two believed unusual animal behavior was a real phenomenon, although 15 responses is too small to really evaluate, given the standard deviation of at least + and - the square root of the number.

Which raises the question: does the ratio of those who believe in the precursor phenomena decrease reciprocally as a function of the distance from the epicenter of any major earthquake?

Reasons for rejection and scepticism:

The following are a few responses to the question, Why have these phenomena not been studied seriously by scientists? (Permission has been given to quote.)

Designing experiments to objectively test these phenomena would be difficult. Getting funding for such a study would probably be even more difficult ...Catherine de Groot-Hedlin.

It is very difficult or impossible to reproduce and/or predict, and therefore to use scientific methodology....Marina Stepanova

Superstition is usually given a wide-berth by the scientific community in this country... Only recently has animal behavior been studied seriously... Many people may not want to test for this for fear that they may need to believe something that is contrary to popular belief based on culture.

Scientists are always behind on natural phenomena. They have to be hit hard on the head before they pay attention. It is easier and safer for them to stay with the conventional beliefs than to risk being the only one wrong (This I feel, falls into the category of pack or mob mentality) ... The few that are not ruled by peer pressure are the ones who make genuine advances in their fields. Most lack the courage to make a stand on something new. Although I do not know this for a fact, I've heard that a fairly accurate prediction of earthquakes can be made by watching the "Lost Pets" section of the newspapers. A higher incidence of missing pets indicates a greater probability of an earthquake of noteworthy size occurring... (Anonymous).

Superstition and Christianity are the two biggest reasons. Fear of the invisible. Also denial of senses and perceptions, which cannot be measured in laboratories. Intuition and such tend to be considered female feelings and since most scientists are men - and scientific education is in the male realm - anything female is discounted and ridiculed, even by women in the scientific field.

Scientists tend to look to the mechanics of the earthquake itself, rather than necessarily examining any of the outlying symptoms that might be present. My feeling is that Western Science is only just beginning to look beyond the boundaries of its rather near-sighted view of phenomenological reality. Western scientists tend to be preoccupied with specific cause and effect, rather than taking a holistic view.

Most animals are highly tuned to their surroundings and possibly use senses based on the earth electromagnetic fields for various purposes and so are more sensitive to distractions. However, I believe some of these may be superstitions based on false observations.

Summary

Although some people believe precursor reports are not to be trusted, enough respondents in different countries were independently saying the same sorts of things that it was worth testing the hypothesis. Not only did we find a general time and location dependence for many of the reported phenomena, but the EM hypothesis also explains phenomena for which other hypotheses have been offered or which have simply been left in the "too hard" basket—such as those explored in Chapter 9.

It seems very reasonable to us that work should continue on developing electromagnetic methods of detecting large scale underground ruptures. After all, they may lead to large earthquakes.

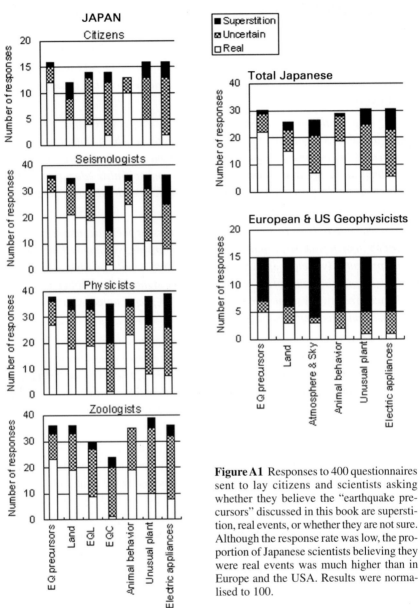

Figure A1 Responses to 400 questionnaires sent to lay citizens and scientists asking whether they believe the "earthquake precursors" discussed in this book are superstition, real events, or whether they are not sure. Although the response rate was low, the proportion of Japanese scientists believing they were real events was much higher than in Europe and the USA. Results were normalised to 100.

Appendix 2

Disaster Prevention

Earthquakes won't wait until science is able to predict them. In earthquake prone areas populations need to be as prepared as possible for a major quake.

A writer in *The Bulletin of the Seismological Society of America* said six lessons had been learned from the San Francisco and Great Kanto (Tokyo-Yokohama quakes) in the 1920s (Jaggar, 1923) and recommended their implementation in earthquake zones. They bear repeating, for the simple reason that they are often followed just after a large quake, but soon neglected again.

- Floors and roofs should be as light as possible, without sacrificing other equally important qualities, such as fireproofing, permanence, etc.
- Walls and columns should be made strong enough to move without collapsing and bringing down the weight of the construction above them.
- Large, open floor areas unsupported by walls and partitions should be avoided.
- Honest, first-class workmanship is of major importance.
- All parts of a building's foundations should be connected together, with connections strong enough to make the various parts act and move together.
- Every roof tile should be individually secured in place so that it cannot slip during an earthquake.

The Shizuoka Disaster Prevention Office in Japan has put together a checklist of emergency goods that should always be to hand in earthquake-prone areas, preferably in a knapsack (Figure A2.1). People can add others.
Other suggestions:

- Remove insecure objects or secure them. Many injuries are caused by unsecured objects falling on people during earthquakes.
- Nail your bookcases and other furniture to the wall to prevent movement.
- Attach movable ornaments to underlying solid surface with adhesives.

These are more sensible precautions to take than moving out of town because a range of unusual phenomena have been observed in the region.

New technologies are being applied to reduce risk in earthquake zones e.g. in New Zealand a dam has been built on a fault line but the large concrete structure has been designed to move in any direction during an earthquake, without water leakage. Many buildings and bridges now employ special lead/rubber combination foundations for prevention of a collapse—though the fate of the rubber during fires caused by quakes is a concern. The search must go on for new technologies of special effectiveness in earthquake zones, though the momentum, unfortunately, slows during seismically quiescent periods—and is almost nonexistent in countries where earthquakes are rarely experienced.

A checklist for the knapsack

- [ ] Helmet
- [ ] Towel & soap
- [ ] Emergency drugs
- [ ] Gloves
- [ ] Flashlight (& extra battery)
- [ ] Cash & ID card
- [ ] Mineral water
- [ ] Matches or lighter & candle

- [ ] Sleeping bag & Blanket
- [ ] Undergarments
- [ ] Tissue paper
- [ ] Rope
- [ ] Portable radio
- [ ] Mobile phone
- [ ] Emergency foods
- [ ] Knife/can-opener

Figure A2.1 Check list of emergency goods to keep in a knapsack in case of earthquake (From a leaflet issued by the Shizoka Prefecture in Japan).

Bibliography and Reading List

Abe N. (1935):   Galvanotropism of the catfish, *Parasilurus asotus* (Linne). *Sci. Rep. Tohoku Univ.* **10**, 393-406.

Adey W.R. (1975) Effects of electromagnetic radiation on the nervous system. *Ann. N.Y. Acad. Sci.* **247**, 15-20.

Aggarwal Y.P., Sykes L.R., Armbruster J. and Sbar M.L. (1973):   Premonitory changes in seismic velocities and prediction of earthquakes. *Nature* **241**, 101-104.

Asano M. (1998): Catfish can sense electricity. *Earthquake Journal* **26**, 52-59 (in Japanese).

Asahara H., Yokoi S., Yamanaka C. and Ikeya M. (2002): An automatic infrared system to observe unusual animal behavior. *J. Atm. Electricity* 22, 223-232.

Barka A. (1999): The 17 August 1999 Izmit Earthquake. *Science.* **285**, 1858-1859.

Barry J.D. (1968): Laboratory ball lightning. *J. Atmos. Terr. Phys.* **30**, 313-317.

Bastian J. (1994):   Electrosensory organisms. *Phys. Today* **47**, 30-37.

Bernardi A., Fraser-Smith A.C., McGill P.R. and Villard O.G. (1991):   ULF magnetic field measurements near the epicenter of the Ms 7.1 Loma Prieta Earthquake. *Phys. Earth Planet. Interiors* **68**, 45-63.

Blanc E. (1985):   Observations in the upper atmosphere of infrasonic waves from natural or artificial source: A summary. *Annal. Geophys.* **3**, 673-688.

Blazquez M.A. and Weigel D. (2000):   Integration of floral inductive signals in Arabidopsis. *Nature* 404, 889-890.

Bose J.C. (1928):   *Motor Mechanism of Plants* (Longmans, Green and Co., London) p.22.

Brady B.T. and Rowell G.A. (1986):   Laboratory investigation of electrodynamics of rock fracture. *Nature* **321**, 488-492.

Bufe C. and Nanevicz J. (1976):   Atmospheric electric field observations, animal behavior, and earthquakes. In *Abnormal Animal Behavior Prior to Earthquakes I,* ed. by Evernden J.F. (USGS, Menlo Park), pp95-106.

Bullock T.H. (1982):   Electroreception. *Ann. Rev. Neurosci.* **5**, 121-170.

Burstein S.G. (1998):   The Alice in Wonderland Syndrome, an update. *Jabberwocky, The Journal of the Lewis Carroll Society* **27**, 23-31.

Buskirk R.E., Everden J.F. and Andriese P.D. ed. (1980):   USGS Proc. Conf. XI: *Abnormal Animal Behavior Prior to Earthquakes II* (USGS, Menlo Park), 237pp.

Buskirk R.E., Frohlich C. and Latham G.V. (1981):   Unusual animal behavior before earthquakes: A review of possible sensory mechanisms. *Rev. Geophys. Space Phys.* **19**, 247-270.

Carroll L. (1962):   *Alice's Adventures in Wonderland and Through the Looking Glass* (Penguin Books, London), 401pp.

Chamberlain J.W. (1995):   *Physics of the Aurora and Airglow* (Am. Geophys. Union, Washington DC), 704pp.

Chen A.T., Ouchi T., Lin A.M., Chen J.C. and Maruyama T. (2000):   Phenomena associated with the 1999 Chi-Chi Earthquake in Taiwan, possible precursors and after effects. *Terr. Atmosph. Ocean. Sci.* **11**, 689-700.

Chow V. H., (1959):   *Open Channel Hydraulics* (McGraw Hill, New York), pp.680.

Chuo Y.J., Liu J.Y., Pulinet S.A. and Chen Y.I. (2002):   The ionosphere perturbation prior to the Chi-Chi and Chia-Yi earthquakes. *J. Geodynamics* **33**, 509-517.

Church C., Burgess D., Doswell C. and Davies-Jones R. eds.(1990):   *The Tornado: Its Structure, Dynamics, Prediction, and Hazards* (Am. Geophys. Union, Washington DC), 637pp.

Cress G.O., Brady B.T. and Rowell G.A. (1987): Sources of electromagnetic radiation from fracture of rock samples in the laboratory. *Geophys.Res. Lett.* **14**, 331-334.

Dai H. (1996): *Great Earthquake can be Predicted*. (Green Arrow Pub. Tokyo), 200pp (in Japanese).

Davies M.S. (1996): Effects of 60 Hz electromagnetic fields on early growth in three plants species and replication of previous results. *Bioelectromagnetics* **17**, 154-161.

Derr J.S. (1973): Earthquake lights: A review of observations and present theories. *Bull. Seismol. Soc. Am.* **63**, 2177-2187.

Derr J.S. (1986): Luminous phenomena and their relationship to rock fracture. *Nature* **321**, 470-471.

Derr J.S. and Persinger M.A. (1986): Luminous phenomena and earthquakes in southern Washington. *Experimentia* **42**, 991-999.

Disso H.W. (1973): A review of analogue model studies of the coast effect. *Phys. Earth Planet. Inter.* **7**, 294-302.

Dinan U. and Yoshino T. (2000): The anomaly unstable polarization fading of HF wave propagation appeared at the earthquake in Turkey in 1999. Abstract of Intern. Wroclaw Symp. on Electromagnetic Compatibility (EMC 2000).

Draganov A.B., Inan U.S. and Taranenko Yu.N. (1991): ULF magnetic signatures at the earth surface due to ground water flow: A possible precursor to earthquakes. *Geophys. Res. Lett.* **18**, 1127-1130.

Du A., Huang Q. and Yang S. (2002): Epicenter fraction by abnormal ULF electromagnetic emissions. *Geophys. Res. Lett.* **29**, 64-1.

Earthquake Research Group of Biophysics Institute of Academia Sinica (1977): *Animals and Earthquakes* (Seismological Press, Beijing).

Efaxias K, Kopanas J., Bogris N., Kapiris P., Antonopoulos G. and Panayiotis V. (2000): Detection of electromagnetic earthquake precursory signals in Greece. *Proc. Japan Acad.* **76**, Ser. **B**, 45-50.

Egawa S. (1991): Catfish and earthquake prediction. *Jishin (Earthquake) Journal* **12**, 8-14 (in Japanese).

Enomoto Y. and Hashimoto H. (1990): Emission of charged particles from indentation fracture of rocks. *Nature* **346**, 641-643.

Enomoto Y. and Hashimoto H.(1994): Anomalous electric signals detected before recent earthquakes in Japan near Tsukuba. In *Electromagnetic Phenomena Related to Earthquake Prediction*, ed. by Hayakawa M. and Fujinawa Y. (Terrapub, Tokyo), pp261-269.

Enomoto Y., Hashimoto H. and Kikuchi K. (1996): Measurements of transient electric current associated with indentation fracture and deformation of brittle solids. *Phil. Mag.* **74**, 1299-1309.

Enomoto Y., Tsutsumi A., Fujinawa Y., Kasahara M. and Hashimoto H. (1997): Candidate precursors: Pulse-like geoelectric signals possibly related to recent seismic activity in Japan. *Geophys. J. Int.* **131**, 485-494.

Enomoto Y. and Zheng Z (1998): Possible evidence of earthquake lightning accompanying the 1995 Kobe earthquake inferred from the Nojima fault gouge. *Geophys. Res. Lett.* **14**, 2721-2724.

Enomoto Y., Asuke F., Zheng Z. and Ishigaki H. (1998): Hardened foliated fault gouge from the Nojima Fault zone at Hirabayashi: Evidence of earthquake lightning accompanying the 1995 Kobe Earthquake? *Island Arc.* **10**, 447-456.

Enomoto Y. (2002): A tornado-type cloud observed on January 9, 1995 prior to the Kobe Earthquake. *Seismoelectromagnetics: Lithosphere-Atmosphere-Ionosphere Coupling*, ed. Hayakawa M. and Molchanov O.A., (TERAPUB, Tokyo) 267-273.

Evernden J.F. ed. (1976): *USGS Proc. Conf. I: Abnormal Animal Behavior Prior to Earthquakes I* (USGS, Menlo Park), 429pp.

Fenoglio M.A., Johnston M.J.S. and Byerlee J.D. (1995): Magnetic and electric fields associated with changes in high pore pressure in fault zones: Application to the Loma Prieta ULF emissions. *J. Geophys. Res.* **100**, 12951-12958.

Feynman R.P., Leighton R.B. and Sands M. (1964): *The Feynman Lectures on Physics* Vol II. Chapter 9. Electricity in the Atmosphere (CIT, 1964) .

Finkelstein D. and Powell J. (1970): Earthquake lightning. *Nature* **228**, 759-760.

Finkelstein D., Hill R. D. and Powell J. (1973): The piezoelectric theory of earthquake lightning. *J. Geophys. Res.* **78**, 992-993.

Fishkova L.M., Gokhberg M.B. and Pilipenko V.A. (1985): Relationship between night airglow and seismic activity. *Annal. Geophys.* **3**, 689-694.

Franzblau E. and Popp C.J. (1989): Nitrogen oxides produced from lightning. *J. Geophys. Res.* **94**, 11089-11104.

Fraser-Smith A.C., Bernardi A., McGrill P.R., Ladd M.E., Helliwell R.A. and Villard Jr. O.G. (1990): Low-frequency magnetic field measurements near the epicenter of the Ms 7.1 Loma Prieta Earthquake. *Geophys. Res. Lett.* **17**, 1465-1468.

Freund F. (2002): Charge generation and propagation in igneous rocks. *J. Geodynamics* **33**, 5432-572.

Freund F. (2002): Seismic electric signals (SES) and the conductivity structure of the crust. *Seismo Electromagnetics: Lithosphere-Atmosphere-Ionosphere Coupling*, 141-152.

Fujinawa Y. and Takahashi K. (1990): Emission of electromagnetic radiation preceding the Ito seismic swarm of 1989. *Nature* **347**, 376-378.

Fujinawa Y., Kumagai T. and Takahashi K. (1992): A study of anomalous underground electric field variations associated with a volcanic eruption. *Geophys. Res. Lett.* **19**, 9-12.

Fujinawa Y., Takahashi K., Matsumoto T. and Kawakami N. (1997): Experiments to locate sources of earthquake-related VLF electromagnetic signals. *Proc. Japan Acad.* **73B**, 33-38.

Fujinawa Y. MatsumotoT. and Takahashi K. (2002): Modeling confined pressure changes inducing anomalous electromagnetic fields related with earthquakes. *J. Appl. Geophys.* **49**, 101-110.

Galli I. (1910): Raccolta e classificazione di fenomeni luminosi osservati nei terremoti. *Bolletino della Societa Italiana* **14**, 221 (in Italian as referred by Musha).

Gawthrop W.H., Johnson R., Haberman R.E. and Wyss M. (1976): Preliminary experiments on behavior of mice before rock failure in laboratory. *Abnormal Animal Behavior Prior to Earthquakes I*, 205-212.

Geller R.J. (1996): VAN: a critical evaluation. In *Critical Review of VAN*, ed. by Lighthill J. (World Scientific, Singapore), pp155-238.

Gokhberg M.B., Morgounov V.A., Yoshino T. and Tomizawa I. (1982): Experimental measurements of electromagnetic emissions possibly related to earthquakes in Japan. *J. Geophys. Res.* **87**, 7824-7828.

Gokhberg M.B., Morgounov V.A. and Pokhotelov O.A. (1995): *Earthquake Prediction: Seismo-electromagnetic Phenomena* (Gordon Breach Pub., Amsterdam), 193pp.

Gutenberg B. and Richter C.F. (1956): Magnitude and energy of earthquakes. *Ann. Geofis.* **9**, 1-15.

Hata M., Takumi I., Yabashi S. and Tian X. (1996): An anomaly of ELF band vertical magnetic flux as a precursor of dome formation at Unzen volcano and its model analysis. *Phys. Earth Interiors*, **105**, 271-277.

Hata M., Takumi I., Yabashi S. and Imaizumi A. (1996): ELF horizontal magnetic field flux precursor of the moderate M5.8 Yamanashi's 96 inland earthquake. *J. Atmos. Electr.* 16, 199-220.

Hatai S. and Abe N. (1932): The responses of the catfish, *Parasilurus asotus*, to earthquakes. *Proc. Imp. Acad. Japan* **8**, 375-378.

Hatai S., Kokubo S. and Abe N. (1932): The earth current in relation to the responses of the catfish. *Proc. Imp. Acad. Japan* **8**, 478-481.

Hayakawa M. and Fujinawa Y. ed. (1994): *Electromagnetic Phenomena Related to Earthquake Prediction* (Terrapub, Tokyo), 677pp.

Hayakawa M., Molchanov O.A., Ondoh T. and Kawai E. (1996): Anomalies in subionospheric VLF signals for the 1995 Hyogo-ken Nambu Earthquake. *J. Phys. Earth* **44**, 413-418.

Hayakawa M. ed. (1999): *Atmospheric and Ionospheric Electromagnetic Phenomena Associated with Earthquakes* (Terrapub, Tokyo).

Hayakawa M. and Molchanov O.A. ed. (2002): *Seismo Electromagnetics: Lithosphere-Atmosphere- Ionosphere Coupling* (Terrapub, 2002) pp477.

Hearn Lafcadio (1897): *"Gleaming in Buddha-Fields : Studies of Hand and Soul in the Far East. (*Houghton, Boston and New York), pp 296.

Hearn Lafcadio (1966): Miminashi-Hoichi; *Classic tale of mystery*; with an introduction of Donald Keene; illustrations by Kuwata M. (Kodansha Intern. Tokyo).

Hendervari P. and Nosziticzius Z. (1985): Recent results concerning earthquake lights. *Ann. Geophys.* **3**, 705-707.

Honkura Y. (1981): Electric and magnetic approach to earthquake prediction, in *Current Research in Earthquake Prediction* I. Ed. Rikitake T. 301-383.

Honkura Y., Matsushima M., Oshiman N., Tuncer M.K., Baris S., Ito Y. and Isikara A.M. (2002): Small electric and magnetic signals observed before the arrival of seismic wave. *Earth Planets Space* **54**, e9-e12, 2002.

Hopkins C.D. (1973): Lightning as background noise for communication among electric fish. *Nature* **242**, 268-270.

Huang Q. and Ikeya M. (1997): Electric field effects on animals: Mechanism of seismic anomalous animal behavior 'SAAB). *Earthq. Res. China* **11**, 109-118.

Huang Q., Ikeya M. and Huang P. (1997): Electric field effects on animals: Mechanism of seismic anomalous animal behaviors (SAABs). *Earthq. Res. China* **11**, 109-118.

Huang Q. and Ikeya M. (1998a): Theoretical investigation of seismic electric field associated with faulting. *Earthq. Res. China* **12**, 295-302.

Huang Q. and Ikeya M. (1998b): Seismic electromagnetic signals (SEMS) explained by a simulation experiment using electromagnetic waves. *Phys. Earth Planet. Inter.* **109**, 107-114.

Huang Q. and Ikeya M. (1999): Experimental study on the propagation of seismic electro-magnetic signals (SEMS) using a mini-geographic model of the Taiwan Strait. *Episodes* **22**, 289-294.

Igarashi G., Saeki S., Takahata K., Sumikawa K., Tasaka S., Sasaki Y., Takahashi M. and Sano Y. (1995): Ground-water radon anomaly before the Kobe Earthquake in Japan. *Science* **269**, 60-61.

Ikeya M., Miki T. and Tanaka K. (1982): Dating of a fault by ESR on intrafault materials. *Science* **215**, 1392-1393.

Ikeya M. (1993; 2002): *New Applications of Electron Spin Resonance - Dating, Dosimetry and Microscopy* (World Scientific, Singapore), 500pp.

Ikeya M. and Takaki T. (1996): Seismic electric signals and animal anomalies. *J. Speleo. Soc. Japan* **20**, 38-47.

Ikeya M. and Takaki T. (1996): Electromagnetic fault for earthquake lightning. *Jpn. J. Appl. Phys.* **35**, L355-L357.

Ikeya M., Takaki S. and Takashimizu D. (1996a): Electric shocks resulting in seismic animal anomalous behaviors (SAABs). *J. Phys. Soc. Japan* **65**, 710-712.

Ikeya M., Furuta H., Kajiwara N. and Anzai H. (1996b): Ground electric field effects on rats and sparrows: Seismic anomalous animal behaviors (SAABs). *Jpn. J. Appl. Phys.* **35**, 4587-4594.

Ikeya M. and Huang Q. (1997): Earthquake frequency and moment magnitude relations for mainshock, foreshocks and aftershocks: Theoretical *b* -values. *Episodes* **20**, 181-184.

Ikeya M. and Matsumoto H. (1997): Reproduced earthquake precursor legends using a Van de Graaff electrostatic generator: Candle flame and dropped nails. *Naturwissenschaften* **84**, 539-541.

Ikeya M., Takaki S., Matsumoto H., Tani A. and Komatsu T. (1997a): Pulsed charge model of fault behavior producing seismic electric signal (SES). *J. Circuits, Systems and Computers* **7**, 153-164.

Ikeya M., Kinoshita Y., Matsumoto H., Takaki S. and Yamanaka C. (1997b): A model experiment of electromagnetic wave propagation over long distances using waveguide terminology. *Jpn J. Appl. Phys.* **36**, L1558-L1561.

Ikeya M., Komatsu T., Kinoshita Y., Teramoto K., Inoue K., Gondou M. and Yamamoto T. (1997c): Pulsed electric fields before Kobe and Izu earthquakes from seismically-induced anomalous animal behavior (SAAB). *Episodes* **20**, 253-260.

Ikeya M., Sasaoka H., Teramoto K. and Huang Q. (1997d): Ferroelectric alignment of piezo-compensating quasi-dipolar charges and formation of tornado-like earthquake cloud. *Ionics* **23**, Suppl.2, 3-11.

Ikeya M (1998): *Why Do Animals Behave Unusually Before Earthquakes*? (NHK Pub., Tokyo), 225pp. in Japanese.

Ikeya M. and Matsumoto H. (1998): Duplicated earthquake precursor anomalies of electric appliances. *South China J. Seismol.* **18**, 53-57.

Ikeya M., Matsuda T. and Yamanaka C. (1998a): Reproduction of mimosa and clock anomalies before earthquakes: Are they "Alice in Wonderland Syndrome"? *Proc. Japan Acad.* **74B**, 60-64.

Ikeya M., Matsumoto H. and Huang Q (1998b): Aligned silkworms as seismic animal anomalous behavior (SAAB) and electromagnetic model of a fault: A theory and laboratory experiment. *Acta Seismol. Sin.* **11**, 365-374.

Ikeya M. (1999): Earthquake precursors due to seismic electromagnetic signals (SEMS). *Recent Res. Devel. Appl. Phys.* **2**, 109-127.

Ikeya M. Yamanaka C., Matsuda T., Sasaoka H., Ochiai H., Huang C., Ohtani N., Komuratani T., Ohta M., Ohno Y. and Nakagawa T.(2000a): Electromagnetic pulses generated by compression of granitic rocks and animal behavior. *Episodes* **21** (2000) 262-265.

Ikeya M., Matsumoto H., Huang Q. and Takaki S. (2000b): Theoretical scaling laws for fault length, seismic electromagnetic signals (SEMS) and maximum appearance area. *J. Earthquake Prediction Research* **8**, 351-360.

Ikeya M., Sato H., Ikuta R. and Yamanaka C. (2002a): Scaling model experiment if electromagnetic waves using earth crust of salt solution. *J. Geodynamics* **33**, 497-508.

Ikeya M., Sato H., Ulusoy U., and Kimura R. (2002b): Split sea and walls of water - Moses' phenomenon at the Izmit Earthquake, Turkey. *Proc. Japan Acad.* **78**, Ser. B, 24-29.

Jackson D.D. and Kagan Y.Y. (1998): The VAN method of earthquake prediction: VAN method lacks validity. *EOS, Trans. Am. Geophys. Un.* **79**, 573.

Jaggar T.A. (1923): The Tokyo-Yokohama Earthquake of Sept. 1, 1923. *Bull. Seismol. Soc. Am.* **13**, 124-146.

Jauchem J.R and Frei M.R. (1995): High-peak power microwave pulses: Effects on heart rate and blood pressure in unanesthetized rats. *Aviation, Space and Environmental Medicine* **66**, 992-997.

Johnston M.J.S. (1997): Review of electric and magnetic fields accompanying seismic and volcanic activity. *Surveys in Geophys.* **18**, 441-475.

Kagita C. (1980): *These are Earthquake Clouds* (Chunichi Press, Nagoya), 186pp (in Japanese).

Kamei Y. (2000): *When an Earthquake occurs - Certainly Catfish knew beforehand* (Sanko-sha, Tokyo), 226pp (in Japanese).

Kameyama K., Kishi Y., Toshimura M., Kanazawa N., Sameshima M. and Tsuchiya T. (2000): Tyrosine phosphorylation in plant bending. *Nature* 407, 37.

Kamogawa M. and Ohtsuki Y.H. (1999): Plasmon model of origin of earthquake related electromagnetic wave noises. *Proc. Japan Acad.,* **75**, Ser. B, 186-189.

Kanamori H. and Anderson D.L. (1975): Theoretical basis of some empirical relations in seismology. *Bull. Seismol. Soc. Am.* **65**, 1073-1096.

Kapitza P.L. (1955): On the nature of ball-lightning. *Dokl. Akad. Nauk. SSSR* **101**, 245-248 (in English); Plasma and the controlled thermonuclear reaction. *Science* **205**, 959-964.

Kapitza P.L. (1969): Ball-lightning and radio emission from linear lightning. *Soviet Physics-Technical Physics* **13**, 1475-1476.

King C.Y. (1984): Evaluating hydrological and geochemical anomalies. *Nature* 312 501-502.

Kirschvink J. L. (2000): Earthquake prediction by animals: Evolution and sensory perception. *Bull. Seism. Soc. Am.* **90**, 312-323.

Kokubo S. (1934): On the behavior of catfish in response to galvanic stimuli. *Sci. Rep. Tohoku Univ.* **9**, 87-96.

Kopytenko Yu. A., Matiashvili T.G., Voronov P.M., Kopytenko E.A and Molchanov O.A. (1993): Detection of ultra-low-frequency emission connected with the Spitak Earthquake and its aftershock activity, based on geomagnetic pulsations data at Dusheti and Vardzia observatories. *Phys. Earth Planet. Inter.* **77**, 85-95.

Kopytenko Yu. A., Ismagilov V., Hayakawa M., Smirnova N., Troyan V. and Peterson T. (2001): Investigation of ULF electromagnetic phenomena related to earthquakes: contemporary achievements and perspectives. *Annali di Geofisica* **44**, 325-334

Kormiltsev W., Kostrov N.P., Ratushnyak A.N. and Shapiro V.A. (2002): *Seismo Electromagnetics: Lithosphere-Atmosphere-Ionosphere Coupling*. 203-207.

Kraemer H.C., Smith B.E. and Levine S. (1976): An animal behavior model for short term earthquake prediction. In *Abnormal Animal Behavior Prior to Earthquakes I*, ed. by Evernden J.F. (USGS, Menlo Park), pp213-232.

Kumazawa M. (1961): Disturbances in electromagnetic field in rocks due to piezoelectric effects in connection with seismic waves. *J. Earth Sci. Nagoya Univ.* **9**, 54-79.

Kusala R., Rajendran C.P., Thakkar M., Tuttle M.P. (2001): *Current Science,* **80,** 1376-1377

Kushida Y. and Kushida R. (2002): Possibility of earthquake forecast by radio observations in the VHF band. *J. Atmospheric Electricity* **22**, no.3, 239-255.

Langmuir I. (1989): Pathological science. *Physics Today* **42**, 36-48.

Lee W.H.K., Ando M. and Kautz W.N. (1976): A summary of literatures on unusual animal behavior before earthquakes. In *Abnormal Animal Behavior Preceding Earthquakes I*, ed. by Evernden J.F. (USGS, Menlo Park), pp15-53.

Lighthill S.J. ed. (1996): *A Critical Review of VAN: Earthquake Prediction from Seismic Electric Signals* (World Scientific, Singapore), 376pp.

Liu J.Y., Chen Y.I., Pulinets S.A., Tsai Y.B. and Cho Y.J. (2000): Seismo-ionospheric signatures prior to M>6.0 Taiwan earthquakes. *Geophys. J. Lett.* 27, 3113-3116.

Lin A., Tanaka N. Uda S.(2001): Infiltration of meteoric and sea water into deep fault zones during episodes of coseismic events: A case study of the Nojima Fault, Japan. *Bull. Earthq. Res. Inst. Univ. Tokyo* **76**, 341-353.

Lockner D.A., Johnston M.J.S. and Byerlee J.D. (1983): A mechanism to explain earthquake light. *Nature* **302**, 28-33

Lott D.F., Hart B.L., Verosub K.L. and Howell M.W. (1979): Is unusual animal behavior observed before earthquakes? YES and NO. *Geophys. Res. Lett.* **6**, 685-687.

Lu Dajong (1981): Observation of earthquake clouds (EQC). *Kexue Tongbao* **26**, 439-442.

Lu Dajong (1990): *Story about Clouds and Earthquakes* (Gaihusha, Tokyo), 149pp (translated into Japanese).

Maeda K. and Tokimasa N. (1996): Decametric radiation at the time of the Hyogo-ken Nanbu Earthquake near Kobe in 1995. *Geophys. Res. Lett.* **23**, 2433-2436.

Matsuda T., Yamanaka C. and Ikeya M. (2001): Behavior of stress-induced charges in cement containing quartz crystals. *Phys. Stat. Sol.* (a) **184**, 359-365.

Matsuda T. and Ikeya M. (2001): Variation of nitric oxide concentration before the Kobe Earthquake, Japan. *Atmospheric Environment* **35**, 3097-3102.

Matsumoto H., Ikeya M. and Yamanaka C. (1998): Analysis of barber-pole color and speckle noises recorded 6 and half hours before the Kobe Earthquake. *Jpn. J. Appl. Phys.* **37**, L1409-L1411.

Matsumoto H., Yamanaka C. and Ikeya M. (2001): ESR analysis of the Nojima fault gouge, Japan from the DPRI 500 m borehole. *Island Arc* **10**, 479-485.

Matteucig G. ed. (1997): *Acts of 4th Int. Conf. Earthq. Precursors* (ICEP, Udine, Italy), 430pp.

McCluskey F.M.J. and Perez A.T. (1992): The electrohydrodynamic plume between a line source of ions and a flat plate: Theory and experiment. *IEEE Trans. Elect. Insul.* **27**, 334-341.

Mead D.F. (1976): Animal behavior preceding the Oroville Earthquake. In *Abnormal Animal Behavior Prior to Earthquakes I*, 55-62.

Medici R.G. (1980): Methods of assaying behavioral changes during exposure to weak electric fields. In *Abnormal Animal Behavior Prior to Earthquakes II*, ed. by Buskirk R.E., Evernden J.F. and Andriese P.D. (USGS, Menlo Park), 114-140.

Milne J. (1888): Note on the effects produced by earthquakes upon the lower animals. *Trans. Seis. Soc. Japan* **12**, 1-4.

Milne J. (1890): Earthquake in connection with electric and magnetic phenomena. *Trans. Seis. Soc. Japan* **15**, 135-162.

Miyakoshi J. (1986): Anomalous time variation of the self potentials in the fractured zone of an active fault preceding earthquake occurrences. *J. Electromagnetic Geoelectric.* **38**, 1015-1030.

Mizuno M., Kashima H., Chiba H., Murakami M. and Asai M. (1998): "Alice in Wonderland" Syndrome as a precursor of depressive disorder. *Psychopathology* **31**, 85-89.

Mizutani H., Ishido T., Yokokura T. and Ohnishi S.(1976): Electrokinetic phenomena associated with earthquakes. *Geophys. Res. Lett.* **3**, 365-368.

Molchanov O.A., Mazhaeva O.A., Goliavin A.N. and Hayakawa M. (1993): Observation by the intercosmos-24 satellite of ELF and VLF electromagnetic emissions associated with earthquakes. *Ann. Geophysics* **11**, 431-440.

Molchanov O.A., Hayakawa M. and Rafalsky V.A. (1994): Penetration of electromagnetic emissions from an underground seismic source into atmosphere, ionosphere and magnetosphere. In *Electromagnetic Phenomena Related to Earthquakes*, ed. by Hayakawa M. and Fujinawa Y. (Terrapub, Tokyo), pp565-606.

Molchanov O.A. and Hayakawa M. (1998): Subionospheric VLF perturbations possibly related to earthquakes. *J. Geophys. Res.* 103, 17,489-17,504.

Moos W.S. (1964): A preliminary report on the effects of electric fields on mice. *Aerosp. Med.* **35**, 374-377.

Morrison H.F. (1976): Electrical phenomena associated with rock strain. In *Abnormal Animal Behavior Prior to Earthquakes I*, ed. by Evernden J.F. (USGS, Menlo Park), pp91-94.

Mulilis J.P. and White M.H. (1986): Behaviors of catfish *Corydoras Aeneus* for use in earthquake prediction. *Earthq. Predict. Res.* **4**, 47-67.

Musha T. (1951): *Japanese Historical Records relevant to Earthquakes* (Mainichi Press, Tokyo), 1019pp (in Japanese).

Musha T. (1957): *Earthquake Catfish* (Toyo-Tosho, Tokyo), 208pp, in Japanese (revised edition was published in 1995, Akashi-Shoten, Tokyo).

Nagao T., Uyeda S., Asai Y. and Kondo Y. (1993): Recently observed anomalous changes in geoelectric potential preceding earthquakes in Japan. *J. Geophys. Res.*

Nagao T. (1998): *New Development of Earthquake Prediction* (Kinmirai-sha, Nagoya), 244pp (in Japanese).

Nagao T., Orihara Y., Yamaguchi T., Takahashi I., Hattori K., Noda Y., Sayaanagi K. and Uyeda S. (2000): Co-seismic geoelectric potential changes observed in Japan. *Geophys. Res. Lett.* **27**, 1535-1535.

Nagao T., Enomoto Y., Fujinawa Y., Hata M., Hayakawa M., Huang Q., Izutsu J., Kushida Y., Maeda K., Oike K., Uyeda S. and Yoshino Y. (2002): Electromagnetic anomalies associated with 1995 Kobe Earthquake. *J. Geodynamics* **33**, 401-411.

T. Nakajima, T. Kunitomo, M. Kumazawa, N Shigeta, H. Nagao (2003): Observations of the transfer function using an electromagnetic sounding system, EM-ACROSS. Abstract of IUGG0 (Sapporo) A.146.

Nikonov A.A. (1992): Abnormal animal behavior as a precursor of the 7 December 1988 Spitak Armenia, earthquake. *Natural Hazards* **6**, 1-10.

Nitsan U. (1977): Electromagnetic emission accompanying fracture of quartz bearing rocks. *Geophys. Res. Lett.* **4**, 333-336.

Ogawa T. and Utada H. (2000): Coseismic piezoelectric effects due to a dislocation 1. An analytic far and early-time field solution in a homogeneous whole space. *Phy. Earth Planet. Interiors* **121**, 273-288.

Ohtsuki Y.H. ed. (1988): *Science of Ball Lightning (Fire Ball)* (World Scientific, Singapore), 339pp.

Ohtsuki Y.H. and Ofuruton H. (1991): Plasma fireballs formed by microwave interference in air. *Nature* **350**, 139-141.

Oike K. and Yamada T. (1994): Relation between shallow earthquakes and electromagnetic noises in the LF and VLF ranges. In *Electromagnetic Phenomena Related to Earthquake Prediction*, ed. by Hayakawa M. and Fujinawa Y. (Terrapub, Tokyo), pp.115-130.

Oike K. and Ogawa T. (1986): Electromagnetic radiation from shallow earthquakes observed in the LF range. *J. Geomag. Geoelectr.*, **38**, 1013-1040.

Omori F.(1923): Pheasant as seismoscope. *Bull. Jpn. Imp. Earthq. Invest. Comm.* **11**, 1-5.

Ondoh T. (1998): Ionosphere disturbances associated with great earthquake of Hokkaido southwest coast, Japan of July 12, 1993. *Phys. Earth. Planet. Inter.* **105**, 261-269.

Ondoh T. and Hayakawa M. (1999): Anomalous occurrence of sporadic E-layers before the Hyogoken-Nambu Earthquake, M7.2 of January 17, 1995. In *Atmospheric and Ionospheric Electromagnetic Phenomena Associated with Earthquakes*, ed. by Hayakawa M. (Terrapub, Tokyo), pp629-639.

Papadopoulos G.A. (1999): Luminous phenomena associated with earthquakes in East Mediterranean. *Atmospheric and Ionospheric Electromagnetic Phenomena Associated with Earthquakes*. Ed. Hayakawa M., 559-575.

Park S.K., Johnston M.J.S., Madden T.R., Morgan F.D. and Morrison H.F. (1993): Electromagnetic precursors to earthquakes in the ULF band: A review of observations and mechanisms. *Rev. Geophys.* **31**, 117-132.

Park S.K (1996): Precursors to earthquakes: Seismoelectromagnetic signals. *Surveys in Geophysics* 17, 493-516.

Parrot M. (2002): The micro satellite DEMETER. *J. Geodynamics* **33,** 535-541.

Parrot M. and Johnson M.J.L. eds (1989): 'Seismoelectromagnetic Effects' *Phys. Earth Planet. Interiors* 1-177.

Payne K.G. and Weinberg F.J. (1958): A preliminary investigation of field-induced ion movement in flame gases and its applications. *Proc. Royal Soc.* **A 250**, 316-336

Petrenko V.F. (1998): Effect of electric field on adhesion of ice to mercury. *J. Appl. Phys.* **84**, 261-267.

Pierce E.T. (1976): Atmospheric electricity and earthquake prediction. *Geophys. Res. Lett.* **3**, 185-188.

Pilipenko V.A., Fedorov E.N., Yagova N.V. and Yumoto K. (1999): Attempt to detect ULF electro-magnetic activity preceding earthquake. In *Atmospheric and Ionospheric Electromagnetic Phenomena Associated with Earthquakes,* ed. by Hayakawa M. (Terrapub, Tokyo), pp203-214.

Pilipenko V.A., Shalimov S., Uyeda S., and Tanaka H. (2001): Possible mechanism of the over-horizon reception of FM radio waves during earthquake preparation period. *Proc. Japan Acad.* **77** Ser. **B**, 125-130.

Rabinovitch A., Frid V. and Bahat D. (1999): A note on the amplitude-frequency relation of electromagnetic radiation pulses induced by material failure. *Phil. Mag. Lett.* **79**, 195-200.

Raleigh B., Benett D., Craig H., Hanks T., Molnat P., Nur A., Savage J. Scholz C., Turner R. and Wu F. (1977): Prediction of Haicheng Earthquake. *EOS*, 58 (5), 236-272.

Rikitake T. (1975): Dilatancy model and empirical formulas for an earthquake area. *Pageoph.* **113**, 141-147.

Rikitake T. (1976): *Earthquake Prediction* (Elsevier, Amsterdam), 357pp.

Rikitake T. (1978): Biosystem behavior as an earthquake precursor. *Tectonophys.* **51**, 1-20.

Rikitake T. (1987): Magnetic and electric signal precursory to earthquake: An analysis of Japanese data. *J. Geomag. Geoelect.* **39**, 47-61.

Rikitake T., Oshiman N. and Hayashi M. (1993): Macro-anomaly and its application to earthquake prediction. *Tectonophys.* **222**, 93-106.

Rikitake T. (1994): Nature of macro-anomaly precursory to an earthquake. *J. Phys. Earth* **42**, 149-163.

Rikitake T. (1998): *Earthquake Prediction and Precursors; Science of Unusual Phenomena before Earthquakes* (Kinmirai-sha, Nagoya), 244pp (in Japanese).

Rikitake T. (2001): *Prediction and Precursors of Major earthquakes* (Terra Sci. Pub., Tokyo) pp197.

Sasaoka H., Yamanaka C. and Ikeya M. (1998): Measurements of electric potential variation by piezoelectricity of granite. *Geophys. Res. Lett.* **25**, 2225-2228.

Sato H., Yamanaka C. and Ikeya M. (2001): Experimental study on the propagation of electromagnetic waves and the spatial distribution of electric potential. Jpn. *J. Appl. Phys.* **40**, 5182-5185.

Schloessin H.H. (1985): Experiments on the electrification and luminescence of minerals and possible origins of EQLs and sferics. *Annal. Geophys.* **3**, 709-720.

Schloessin H.H. (1994): Mechanisms and electrification preceding earthquakes. In *Electromagnetic Phenomena Related to Earthquake Prediction,* ed. by Hayakawa M. and Fujinawa Y. (Terrapub, Tokyo), 495-510.

Scholz C.H., Sykes L.R. and Aggarwal Y.P. (1973): Earthquake Prediction: A physical basis. *Science* **181**, 803-809.

Semm P., Schneider T. and Vollrath L. (1980): Effects of an Earth-strength magnetic field on electric activity of pineal cells. *Nature* **288**, 607-608.

Serebryakova O.N., Bilichenko S.V., Chmyrev V.M., Parrot M., Rauch J.L., Lefeuvre F. and Pokhotelov O.A. (1992): Electromagnetic ELF radiation from earthquake regions as observed by low-altitude satellites. *Geophys. Res. Lett.* **19**, 91-94.

Sheldrake R. (1994): *Seven Experiments That Could Change the World: A Do-It-Yourself Guide to Revolutionary Science* (Fourth Estate, London), 269pp.

Shimada Y., Uchida S., Yasuda Y., Motokoshi S., Yamanaka C., Kawasaki Z-I., Yamanaka T., Ishikubo A. and Adachi Y.(1997): Laser triggered lightning. *Proc. SPIE Conference on High Power Laser Ablation*, **3423**, 258-261.

Silver P.G. and Wakita H. (1996): A search for earthquake precursors. *Science* **273**, 77-79.

Skidmore W.D. and Baum S.J. (1974): Biological effects in rodents exposed to 108 pulses of electromagnetic radiation. *Health Phys.* **26**, 391-398.

Slifkin L.M. (1993): Seismic electric signals from displacement of charged dislocations. *Tectonophys.* **224**, 149-154.

Smith S.D., McLeod B.R. and Liboff A.R. (1993): Effect of CR-tuned 60 Hz magnetic fields on sprouting and early growth of *Raphanus sativus. Bioelectrochem. Bioenerg.* **32**, 67-76.

Soler J.M. (2000): One-dimensional reactive transport modelling of the interaction between a high-pH plume and a fractured granodiorite - the GTS-HPF Project. *J. Conf. Abstracts.* **5**, 947.

Staelin D.H., Morgenthaler A.W. and Kong J.A. (1994): *Electromagnetic Waves* (Prentice Hall, New York), 562pp.

Stasko A.B. (1971): Review of field studies on fish orientation. *Ann. N.Y. Acad. Sci.* **188**, 12-29.

Sternheim M.M. and Kane J.W. (1986): *General Physics* (John Wiley & Sons, New York), 765pp.

Sugihara M., Funakoshi H., Yamamoto N., Ichida S., Matsuda S., Washio K. and Hataya M. (1998): *The Animal Rescue at the Great Hanshin Earthquake* (Trais Co., Kobe), 114pp (in Japanese).

Suyehiro Y. (1934): Some observations on unusual behavior of fishes prior to an earthquake. *Bull. Earthq. Res. Inst. Tokyo Univ.* Suppl.**1**, 228-231.

Takahashi K. and Fujinawa Y. (1993): Locating source regions of precursory seismoelectric fields and mechanism generating electric field variations. *Phys. Earth Planet. Inter.* **77** 33-38.

Takaki S. and Ikeya M. (1998): A dark discharge model of earthquake light. *Jpn. J. Appl. Phys.* **37**, 5016-5020.

Terada T. (1931): On luminous phenomena accompanying earthquakes. *Bull. Earthq. Res. Inst. Tokyo Univ.* **9**, 225-255.

Terada T. (1932): On some probable influence of earthquakes upon fisheries. *Bull. Earthq. Res. Inst. Tokyo Univ.* **10**, 393-401.

Teramoto K. and Ikeya, M. (2000): Experimental study of cloud formation by intense electric fields. *Jpn. J. Appl. Phys.* **39**, 2876-2881.

Toriyama H. (1966): The behavior of the sensitive plant in a typhoon. *Bot. Mag. Tokyo* **79**, 427-428.

Toriyama H. and Jaffe M.J. (1972): Migration of calcium and its role in the regulation of seismonasty in the motor cell of *Mimosa pudica* L. *Plant Physiol.* **49**, 72-81.

Toriyama H. (1991): Individuality in the anomalous bioelectric potential of silk trees prior to earthquakes. *Sci. Reports Tokyo Woman's College* 90, 1067-1077.

Toutain J.P. and Baubron J.C. (1999): Gas geochemistry and seismotectonics: a review. *Tectonophysics* **304**, 1-27.

Tributsch H. (1978): Do aerosol anomalies precede earthquake? *Nature* **276**, 606-608.

Tributsch H. (1982): *When the Snakes Awake* (MIT Press, Cambridge), 248pp.

Tripp H.M., Warman, G.R. and Arendt, J. (2003): Circularly polarised MF (50 Micro-T 50 Hz) does not acutely suppress melatonin secretion from cultured Wistar rat pineal glands. *Bioelectromagnetics* 24, 118-124.

Tronin A.A., Hayakawa M. and Molchanov O.A. (2002): Thermal IR satellite data applications for earthquake research in Japan and China. *J. Geodynamics* **33**, 519-534.

Tsukuda T. (1997): Sizes and some features of luminous sources associated with the 1995 Hyogo-ken Nanbu Earthquake. *J. Phys. Earth* (Japan) **45**, 73-82.

Tsunogai U. and Wakita H. (1995): Precursory chemical changes in ground water: Kobe Earthquake, Japan. *Science* **269**, 61-63.

Tsutsui M. (2002): Detection of earth-origin electric pulses. *Geophys. Res. Lett.* **29**, (8) 35/1-4.

Tzanis A. and Vallianatos F. (2001): A critical review of electric earthquake precursors. *Annali di Geofisica* 44, 429-460.

Tzanis A. and Vallianatos F. (2001): A physical model of electric earthquake precursors due to crack propagation and motion of charged edge dislocations. *Seismo Electromagnetics* ed. Hayakawa M. and Molchnov OA. 117-130.

Uda S., Lin A. and Takemura K. (2001): Crack-filling clays and weathered cracks in DPRI 1800 m core near the Nojima Fault, Japan: Evidence for deep surface-water circulation near an active fault. *Island Arc* **10**, 439-446.

Ulusoy U. and Ikeya M. (2001): *Precursor Statements of Earthquakes* (Neyir Publisher, Ankara, 2001) 297pp (in Turkish).

Ulusoy U. and Ikeya M. (2004): Retrospective statements of unusual phenomena before the Izmit Earthquake in 1999. *Bull. Earthquake Res. Inst. Univ. Tokyo* (Submitted).

Utada H. (1993): On the physical background of the VAN earthquake prediction method. *Tectonophysics* **224**, 149-152.

Uyeda S. (1996): Introduction to the VAN method of earthquake prediction. In *A Critical Review of VAN,* ed. by Lighthill J. (World Scientific, Singapore), pp. 3-28.

Uyeda S. (1998): The VAN method of earthquake prediction: VAN method of short-term earthquake prediction shows promise. *Eos, Trans. Am. Geophys. Un.* **79**, 573.

Uyeda S., Hayakawa M., Nagao T., Molchanov O., Hattori K., Orihara Y., Gotoh K., Akinaga Y. and Tanaka H. (2002): Electric and magnetic phenomena observed before the volcano-seismic activity in 2000 in the Izu Island Region, Japan. *PNAS* **99**, 7352-7355.

Varotsos P. and Alexopoulos K. (1986): Stimulated current emission in the earth : piezoelectric currents and related geophysical aspects. In *Thermodynamics of point defects and their relation with bulk properties,* ed. by Amelinckx S., Gevers R. and Nihoul J. (North-Holland, Amsterdam).

Varotsos P., Lazaridou M., Eftaxias K., Antonopoulos G., Makris J. and Kopanas J. (1996): Short term earthquake prediction in Greece by seismic electric signals. In *A Critical Review of VAN,* ed. by Lighthill J. (World Scientific, Singapore), pp29-76.

Volarovich M.P. and Sobolev G.A. (1965): Use of the piezoelectric effects of rocks for subsurface exploration of piezoelectric media. *Dokl. Akad. Nauk. SSSR* **162**, 11-13.

von Hentig H. (1923): Reactions of animals to change in physical environment I: Animal and earthquake. *J. Comp. Psychol.* **3**, 61-71.

Wadatsumi K. ed. (1995): *1519 Statements on Precursors* (Tokyo Pub, Tokyo), 265pp (in Japanese).

Wadatsumi K. (1998): *Precursors for great earthquakes* (Kawade-Shobou, Tokyo), 208pp (in Japanese).

Walcott C. and Green R.P. (1974): Orientation of homing pigeons altered by a change in the direction of an applied magnetic field. *Science* **184**, 180-182.

Walker M.M., Diebel C.E., Haugh C.V., Pankhurst P.M., Montgomery J.C. and Green C.R. (1997): Structure and function of the vertebrate magnetic sense. *Nature* **390**, 371-376.

Walton A. J. (1977): Triboluminescence. *Adv. Phys.* **26**, 887-948.

Wang D., Ushio T., Kawasaki Z-I., Matsura K., Shimada Y., Uchida S., Yamanaka C., Izawa Y., Song Y. and Shimokura N. (1995): A possible way to trigger lightning using a laser. *J. Atmos. Terr. Phys.* **57**, 459-466.

Warwick J.W., Stoker C. and Meyer T.R. (1982): Radio emission associated with rock fracture: Possible application to the great Chilean Earthquake on May 22, 1960. *J. Geophys. Res.* **87**, 2851-2859.

Whitehead N.E., Ikeya M., and Ulusoy U. (2003): The importance of statistical analysis in earthquake prediction. *The 1st International Workshop on Earthquake Prediction, Athens, Greece.* November 6-7, 2003.

Wilson C.T.R. (1897): Condensation of water vapor in the presence of dust-free air and other gases. *Philos. Trans.* **189**, 265-305.

Wilson E.K. (1986): 60-Hz electric field effects on pineal melatonin rhythms: Time course for onset and recovery. *Bioelectromagnetics* **7**, 239-242.

Wyss M. (1997): Second round of evaluations of proposed earthquake precursors. *Pure appl. geophys.* **149**, 3-16.

Yagi T. (1999): Molecular mechanisms of Fyn-tyrosine kinase for regulating mammalian behaviors and ethanol sensitivity. *Biochem. Pharmacology* **57**, 845-850.

Yalciner A.C. and Altinok Y. (2000): Kocaeli Turkish Earthquake of August 17, 1999 Reconnaissance report. *Earthquake Spectra* **16** (Supp. A) 55-62.

Yamada T. and Oike K. (1996): Electromagnetic radiation phenomena before and after the 1995 Hyougo-ken Nambu Earthquake. *J. Phys. Earth* **44**, 405-412.

Yamanaka C., Asahara H., Natsunoto H. and Ikeya M. (2002): Wideband environmental electromagnetic wave observation searching for seismo-electromagnetic signals aand simultaneous observation of catfish behavior - The case for the Western Tottori and the Geiyo Earthquakes-. *J. Atm. Electricity* **22**, 277-290.

Yamanaka T., Uchida S., Shimada Y., Yasuda Y., Motokoshi S., Tsubakimoto K., Kawasaki Z., Ishikubo A, Adachi Y. and Yamanaka C. (1998): First observation of laser trig-gered lightning. *Proc. SPIE Conference on High Power Laser Ablation*, **3343**, 281-288.

Yasui T. (1968): A study of luminous phenomena accompanied with earthquake (Part I). *Mem. Kakioka Mag. Obs.* **13**, 25-61 (in Japanese).

Yasui T. (1971): A study of luminous phenomena accompanied with earthquake (Part II). *Mem. Kakioka Mag. Obs.* **14**, 67-78 (in Japanese).

Yokoi S. Ikeya N., Yagi S., Nagai K. (2002): Mouse circadian rhythm before the Kobe Earthquake and effects of exposure to electromagnetic pulses. *Material Integration* *15*, 59-61.

Yokoi S., Ikeya N., Yagi S. and Nagai K. (2003): Mouse circadian rhythm before the Kobe Earthquake in 1995. *J. Bioelectromagnetics* **24**, 289-291.

Yoshida S., Manjgaladze P., Zilpimiani D., Ohnaka M. and Nakatani M. (1994): Electro-magnetic emissions associated with frictional sliding rock. In Hayakawa M. and Fujinawa Y. ed. (1994), 304-322.

Yoshida S., Clint O.C. and Sammonds P.R. (1998): Electrical potential changes prior to shear fracture in dry and saturated rocks. *Geophys. Res. Lett.* **25**, 1577-1580.

Yoshino T. (1991): Low frequency seismogenic electromagnetic emissions as precursors to earthquakes and volcanic eruptions in Japan. *J. Sci. Explor.* **5**. 121-144.

Yoshino T. and Sato H. (1998): The experimental results on actual measurement of energy transmission loss of magnetic field component across the tunnel. *Phys. Earth Planet. Interiors* **105**, 287-295.

Yue X. and Lu D. (1988): Elementary analysis of the "Interpretation of Earthquakes" Long Huamin. In *Proceedings of the 3rd International Conference on Earthquake Prediction*, ed. by Matteucig G. , pp403-406.

Zhao D., Kanamori H., Negishi H and Wiens D (1996): Tomography of the source area of the 1995 Kobe Earthquake: Evidence for fluids at the hypocenter? *Science* **274**, 1891-1894.

Zlotnicki J. and Le Mouel J.L. (1990): Possible electrokinetic origin of large magnetic variations at La Fournaise volcano. *Nature,* **343***,* 633-636.

Index